The Zero Epoch Guidebook: Cyber Risk, Quantum Threats and Trust Recovery

Dr. James W. Howell, Jr.

A SMARTCIO, LLC BOOK   NEVADA

The Zero Epoch Guidebook:
Cyber Risk, Quantum
Disruption and the Future of
Trust

Copyright © 2026 Dr. James W. Howell, Jr.
All rights reserved.

No part of this book may be reproduced, stored in a retrieval system, or transmitted in any form or by any means, electronic, mechanical, photocopying, recording, or otherwise, without prior written permission of the publisher, except for brief quotations used in reviews or scholarly works.

First edition 2026

Published by SmartCIO, LLC
304 S. Jones Blvd.
Las Vegas, NV 89107
https://www.smartcio.com

ISBN 979-8-9943826-0-8 paperback
ISBN 979-8-9943826-1-5 hardcover
ISBN 979-8-9943826-2-2 ebook
Library of Congress Control Number: 2026928303

Printed in the United States of America

Disclaimer.
The views expressed in this book are those of the author and do not necessarily reflect the views of any organization or institution. This book is for informational purposes only and does not constitute legal, financial, or professional advice.

Dedication

To every CIO, CISO, and IT Director who wakes before dawn to read another incident report, patch another flaw, or calm another storm. Your vigilance protects more than data. It preserves the quiet faith that systems will work, that hospitals will heal, that lights will stay on.

You fight battles most will never see and resolve crises that disappear before they are known. You lead not through fear, but through faith, in your teams, your mission, and the belief that truth can still be trusted.

This book honors your work, your endurance, and your conviction that technology, guided by integrity, can still serve humanity.

Table of Contents

Introduction 1

Chapter 1: Midnight at the Edge of Certainty 4

Chapter 2: The Fragility of Digital Trust 10

Chapter 3: When Math Stops Working 24

Chapter 4: When Machines Learn to Lie 44

Chapter 5: When Governance Cannot Keep Pace 59

Chapter 6: When Intelligence and Power Converge 74

Chapter 7: The Supply Chain of Doubt 84

Chapter 8: Post-quantum Cryptography Made Simple 96

Chapter 9: The Day Trust Breaks 111

Chapter 10: Rebuilding Trust 122

Epilogue: Final Reflection 137

Appendix: Zero Epoch Checklist 148

Introduction

"The next global crisis will not begin with a weapon or a virus. It will begin with doubt."
- Dr. James Howell

The world is entering a moment that feels quiet on the surface yet seismic underneath. Most cyber professionals sense something shifting. They notice strange stories in the news, unusual alerts in their inbox, and increasing warnings from companies, governments, and security experts. They feel the ground move but cannot name the fault line. This book gives that fault line a name: The Zero Epoch.

The Zero Epoch begins when quantum computing can break RSA in a practical, repeatable, and scalable way, ending asymmetric cryptographic assurance even as digital systems continue to operate normally. These conditions are defined in Chapter 3. It is the tipping point where old assumptions collapse and new rules emerge. It marks the transition from an internet held together by mathematics, institutions, and regulations into a world where those anchors no longer guarantee safety.

The Zero Epoch is not a single event. It is a shift in the human relationship with technology, truth, identity, and security. We do not know whether that moment is five years away or thirty. But when midnight of the Zero Epoch strikes, its consequences will be met with global trepidation, fueled by an incredible sense of urgency.

Why Trust Is Breaking Now

Digital trust has always been invisible yet indispensable. You cannot see the algorithms behind your banking app, your medical chart, or social media.

But you rely on them every day. They are the locks and keys of modern civilization.

Those locks are already threatened. Certificate authorities have been breached. Supply chains have been poisoned. Ransomware has paralyzed hospitals. Expired certificates have crashed platforms. These failures show that digital trust is more fragile than it appears. We'll discuss examples more in-depth.

Quantum computing threatens to shatter the locks completely. But it will not act alone. Artificial intelligence is becoming a master forger, able to mimic any voice, image, or identity. Governance, regulatory compliance, and policy, once thought of as the framework of guardrails, now trail far behind.

At the end of the last century, the Y2K event proved that humanity could rally around a defined technical threat when a clear countdown exists. However, the Zero Epoch offers no such clarity. It has no date, no patch, and no finish line. Only inevitability.

Who This Book Is For

This book is not a textbook, though it teaches. It is not a manifesto, though it calls for action. It is a field guide that serves as a map, a survival kit, and an invitation for three audiences that must prepare together.

- **IT consumers:** Gain awareness of how daily life depends on fragile systems, why those systems may fail, and what you can expect if we do nothing to prepare.
- **Students and researchers:** Explore best practices, case studies, and models that provoke debate and inspire design.
- **Professionals and leaders:** Enforce cybersecurity frameworks, harden systems, and use the Zero

Epoch Readiness Checklist to identify gaps in organizational readiness.

Reflection

The Zero Epoch is not about a single technology. It is about convergence: quantum disruption, AI deception, and governance paralysis. Adversaries are already breaking into systems and stockpiling encrypted data, forging synthetic identities, and exploiting compliance gaps. The question is not whether the locks will break. It is whether we will be ready when they do.

Every technological revolution begins with faith. We trust our machines, our networks, and our encryption to protect us. We build entire economies and governments on the assumption that digital trust will hold.

But assumptions are not defenses. The next disruption will not come from malware or espionage. It will come from mathematics itself. Quantum computing will turn encryption from protection into illusion. Artificial intelligence will make deception scalable. Trust, once fractured, will not return easily.

The Zero Epoch is inevitable, but disaster is not. One outcome is a collapse born of complacency, shaped by delay and disbelief. The other is renewal through readiness shaped by preparation and clarity. Midnight is coming. The only question is whether we greet it proactively, prepared with awareness, continuity, and structure, or meet midnight completely in the dark, unprepared, wading through uncertainty, misinformation, and identity loss while watching systems fail. The next era will be shaped by choices made before the disruption, not after.

Chapter 1: Midnight at the Edge of Certainty

The Threshold We Did Not Prepare For

The digital world grew on the assumption that trust could scale indefinitely. Certificates would authenticate. Encryption would protect digital transactions. Identities would remain stable. People believed that mathematics would anchor truth in place. For decades, that belief held. Each new technology is layered on top of the old without challenging the foundation. The internet expanded. Cloud systems accelerated. AI emerged with confidence but without restraint. The pace felt ambitious but manageable.

That confidence hid a structural weakness. The systems that power modern civilization were built for a slower age. They assumed that computation would evolve predictably, that adversaries would follow rules, and that cryptography would last for decades. They did not anticipate a world where quantum acceleration undermines the assumptions that protect global networks. They did not anticipate artificial intelligence capable of imitating authority, rewriting perception, or manufacturing trust out of thin air. The Zero Epoch is the point at which those assumptions fail, and the world realizes that its digital spine cannot support its own weight.

Midnight in the Zero Epoch is not a synchronized global event or a single moment on the clock. It is a perceptual threshold, the point at which enough systems fail quietly and independently that confidence collapses everywhere without a clear beginning or a shared alarm.

The Slow Unraveling of Certainty

Nothing catastrophic announces the start of the Zero Epoch. Instead, small failures begin to cluster. A bank reports a minor verification delay. A government website loads with a security warning. A hospital receives diagnostic results that do not match previous baselines. Each issue is resolved quickly, and each resolution strengthens the illusion that the system is fine. The pattern continues until the failures no longer feel random.

Complexity becomes the adversary. Organizations operate on stacks of dependencies they no longer understand. Certificates issued years ago continue to validate systems that have changed dozens of times. AI-generated forgeries circulate so convincingly that even trained analysts don't hesitate. People tell themselves that everything is under control. They assume stability because they need to believe it. Meanwhile, adversaries collect encrypted data, confident that future machines will expose it.

The Zero Epoch does not begin with a breach. It begins with a belief that a breach is impossible.

When Intelligence Learns to Deceive

Artificial intelligence once served as a tool. It answered questions, summarized documents, and detected anomalies. Today, it fabricates authority with ease. It creates synthetic voices that mimic executives. It generates images that bypass verification. It produces content that aligns precisely with human bias. In this landscape, attackers do not guess what people fear. They calculate it.

Quantum computation accelerates this problem. When data becomes transparent to those who can break its encryption, trust becomes a contested resource. A forged signature, once a rarity reserved for elite adversaries, becomes trivial. A false certificate, once detectable, becomes indistinguishable from the real one. The combination of AI and quantum capability turns deception into a scalable service. The world grows more connected, yet less verifiable.

The World Approaching Midnight

The Zero Epoch approaches quietly. It does not feel cinematic at first. It feels slightly inconvenient. A login takes longer. A security update loops. A website that was safe yesterday now displays an uncertain identity. People refresh the page and assume the network is slow. Engineers write quick patches and assume the issue is routine. Leaders authorize minor investigations without realizing they are looking at the first signs of systemic fatigue.

These hesitations become the soundtrack of a world drifting toward midnight. Trust erodes without the drama of collapse. Systems continue to function, but the confidence beneath them weakens. Once this drift begins, recovery becomes more difficult because each minor failure teaches users to doubt what they see. By the time the warnings become clear, the foundations have already shifted. Midnight is not a moment of impact. It is the moment the world recognizes that the impact has been happening for years.

The Moment Trust Becomes Fragile

Trust is the expectation that a system will behave as intended, even when we are not watching it. It is the quiet confidence that messages are authentic, records are accurate, and actions taken in one place will be honored in another. In the digital world, trust allows speed. It removes the need to question every transaction, every instruction, and every identity.

That confidence fails faster than technology itself. People begin to question whether an email is genuine. They hesitate before approving transactions. They reread official messages that once required no thought. What starts as individual caution spreads outward. Banks shift to manual verification. Hospitals double-check patient data. Governments request physical confirmation of records that were previously accepted without challenge. The world slows because certainty has fractured.

Once digital trust becomes fragile, every system becomes suspect. A valid-looking certificate no longer reassures. A digital signature no longer proves authorship. A routine update could be maintenance, or it could be the mechanism that compromises an entire supply chain. Organizations discover that they cannot function when every interaction requires proof at every step.

The Zero Epoch exposes a fundamental truth. Trust is not created by technology alone. It lives in human confidence. Cryptography supports trust, but it does not replace belief. When belief is shaken, efficiency collapses, coordination degrades, and systems designed for speed revert to caution. Trust, once damaged, does not return automatically. It must be rebuilt deliberately, visibly, and over time.

Why This Moment Matters

The Zero Epoch is not a prediction. It is a trajectory. It represents the collision between accelerating capability and lagging preparation. Societies depend on digital trust for everything from healthcare to national defense. The collapse of that trust, even temporarily, would disrupt daily life for billions of people. The purpose of naming the Zero Epoch is not to spread fear. It is to articulate the stakes. If humanity recognizes the vulnerabilities early enough, it can reinforce the foundations before they crack. If it ignores them, the collapse will be quiet, glacial, and in hindsight, preventable.

This era is not defined by quantum machines or AI models alone. It is shaped by the choices people make in response to them. Leadership, governance, engineering, and awareness will decide how the Zero Epoch unfolds. Technology moves quickly. Responsibility must move faster.

Reflection

Every civilization faces a moment when its assumptions no longer match its reality. The Zero Epoch represents that moment for the digital age. It signals a turning point where the systems people trust begin to behave in unfamiliar ways and the protections once considered permanent start to weaken. It does not call for panic. It calls for clarity.

The threat is not a sudden collapse. It is the slow realization that long-held beliefs about identity, privacy, secrecy, and truth no longer fit the environment we live in. Societies must confront this fragility before it becomes visible in daily life. They must accept that the digital world depends on math, perception, and governance that were designed for conditions that no longer exist.

Midnight has not arrived, but the clock continues its quiet progression. The warning signs appear in scattered incidents, small failures, and subtle shifts in trust. These early signals prepare readers for the chapters ahead, where quantum computing challenges the foundations of cryptography, artificial intelligence reshapes perception, and governance struggles to keep pace with the speed of change.

What happens next depends on whether people recognize the clock ticking and choose to act before the hour strikes. The Zero Epoch is not destiny. It is a test of awareness, leadership, and preparation. The chapters that follow explore that test from every angle, revealing the pressures that are building and the decisions that will determine the future of digital trust.

Chapter 2: The Fragility of Digital Trust

The Architecture Beneath Every Click

Digital trust is the invisible foundation of modern life. Every login, phone text, wire transfer, medical record, email, social media post, and cloud transaction depends on it. We rarely think about this foundation because it rarely asks for attention. The system feels automatic. It feels effortless. It feels inevitable. But trust is not automatic. It is constructed, layered, and maintained by systems that can falter.

    Digital trust rests on three proofs. They seem simple in theory, but they support almost everything you do online in your daily lives. These proofs are also referred to as the Cybersecurity Triad or CIA.

Proof of Confidentiality: Encryption protects your privacy. It turns your data into unreadable code, so only the intended recipient can unlock it. When you check your bank account, send medical information, or message a coworker, encryption keeps outsiders from seeing the details. Without this proof, every transaction would be exposed.

Proof of Integrity: Hash functions and certificates ensure that data has not been changed. They allow systems to detect even the slightest alteration. When you download a file or connect to a secure website, integrity checks confirm that nothing was modified in transit. They protect you from external tampering and silent corruption.

Proof of Authenticity: Digital signatures prove that a message or file came from the person who claims to have sent it. They act like a sealed envelope. If the signature matches, the receiver knows the message is genuine. When you update an app, sign a document, or trust a software patch, you rely on this proof to confirm the sender's identity.

These proofs operate quietly beneath the surface. Most users never see the math behind them. They see the symbols instead: the lock icon in the browser, the two-factor prompt, the green bar of a valid certificate. These signals create a sense of safety even when the infrastructure behind them may be strained or outdated.

The math behind these proofs has remained strong. The weakness often appears in the surrounding system. Encryption, signatures, and integrity checks rely on careful implementation. They also depend on the organizations that manage keys, configure servers, issue certificates, and update software. When those tasks are rushed, ignored, or delegated without oversight, the protection weakens even if the algorithms remain sound.

The architecture of trust depends on human decisions at every level. Engineers decide how keys are stored. Administrators choose when updates are applied. Vendors decide how their software is built and who can access their code. Governments determine which standards to enforce and how often to review them. Each decision shapes the system's strength.

Any weak link becomes a point of failure for everyone who relies on it. A certificate authority with poor security can expose millions of users. A vendor with a vulnerable update process can compromise entire supply chains. A misconfigured server can bypass encryption and expose private data. The failure does not stay local. It travels across networks, services, and organizations.

Digital trust is only as strong as the least careful participant in the chain. This is why trust collapses quietly. The math stands firm while the systems around it falter through oversight gaps, rushed deployments, expired certificates, or outdated hardware. The danger emerges not from a single mistake but from a collection of minor weaknesses that align at the wrong moment.

Why Digital Identity Is Brittle

Digital identity is the core of trust. It answers the most basic question in the digital world. Who are you?

In this book, identity infrastructure refers to the combined systems that establish, verify, and sustain digital identity, including cryptographic keys, certificates, authentication services, governance processes, and the human decisions that maintain them over time.

Chapter 2: The Fragility of Digital Trust

In theory, the answer is straightforward. A certificate authority verifies identity. A digital certificate confirms it. A signature proves that an action came from the right source. A token grants access. Together, these mechanisms allow systems to trust each other without human involvement.

A Certificate Authority, or CA, is an institution that issues digital certificates after verifying identity. It acts as a trusted referee. When a CA vouches for a website, a server, or a piece of software, every device that trusts that authority automatically accepts the claim. This process happens billions of times a day without users ever seeing it.

In practice, this trust rests on institutions that few people recognize and even fewer scrutinize. A small number of CAs anchor the identity of most of the internet. Operating systems and browsers embed their root certificates by default. This design makes digital interaction fast and frictionless. It also concentrates risk. If one trusted authority is compromised or mismanaged, impersonation becomes possible on a global scale. A forged certificate is not a minor inconvenience. It is a skeleton key.

Public Key Infrastructure, or PKI, is the framework that makes this system work. It governs how certificates are issued, validated, trusted, and revoked. PKI allows devices, applications, and people to recognize each other as legitimate without prior contact. It enables browsers to trust websites, software to trust updates, and systems to exchange information securely at scale. Without PKI, every digital interaction would require manual verification.

PKI operates through a hierarchy of trust. At the top sit root CAs whose legitimacy is accepted by operating systems and platforms. Beneath them flow millions of certificates that authenticate servers, users, and devices. This hierarchy supports speed and scale, but it also amplifies failure. Billions of systems depend on a small number of trusted roots.

That concentration creates fragility. When a central certificate authority is breached, misconfigured, or coerced, confidence erodes far beyond a single organization. A failure at the root cascades downward. Certificates that appear valid become suspect. Secure connections feel uncertain. Trust spreads quickly when it is healthy. It collapses just as quickly when its foundation is shaken.

PKI rarely draws attention because it works quietly in the background. Users do not think about certificates when they open a browser or install an update. The Zero Epoch forces that invisibility into view. When the assumptions behind PKI no longer hold, digital trust stops being automatic and becomes something that must be questioned, verified, and rebuilt deliberately.

Certificates also expire. They can be mis-issued. Revocation systems often fail silently. Most users rely on a lock icon that signals encryption, not trustworthiness. Attackers understand this distinction. They obtain valid certificates for malicious domains and present themselves as legitimate. Users see the lock and assume it is safe. This misunderstanding is not a minor flaw. It is the opening that attackers exploit every day.

Case Studies in Trust Failure

DigiNotar: The Collapse of a Guardian (2011)

In 2011, attackers infiltrated DigiNotar, a Dutch CA. They created more than 500 fraudulent certificates for major platforms, including Google, Yahoo, Mozilla, and Skype. During the attack, about 300,000 Iranian users visited counterfeit Google domains. Their credentials and private messages were captured without their knowledge.

When the breach became public, every major browser revoked trust in DigiNotar certificates. Dutch government services lost authentication and faced weeks of outages. Emergency migration efforts cost millions of euros. DigiNotar collapsed within two months. One compromised authority triggered a global identity failure because the trust model had no fallback once its anchor failed.

Impact:
- Users affected: About 300,000
- Fraudulent certificates: More than 500
- Financial cost: Millions of euros
- Downtime: Weeks across European services
- Outcome: Company liquidation
- Trust lesson: A single certificate authority can compromise entire regions

Stuxnet: Weaponized Authenticity (2010)

In 2010, the Stuxnet worm spread by using two stolen certificates from RealTek and JMicron. These certificates made the malware appear legitimate. Windows systems trusted the code because the signatures matched known vendors. The worm then infiltrated Iranian nuclear facilities and targeted the Natanz enrichment site.

Stuxnet damaged about 1,000 centrifuges and disrupted operations for months. No cryptographic weakness was needed. The system failed because a forged identity delivered harmful instructions that appeared legitimate. Stuxnet proved that a valid signature can enable physical damage on a scale typically associated with military operations.

Impact:
- Certificates stolen: 2
- Physical equipment destroyed: About 1,000 centrifuges
- Financial cost: Hundreds of millions
- Downtime: Months of operational delays
- Trust lesson: Trusted signatures can be turned into strategic weapons

Equifax: Identity at National Scale (2017)

In 2017, Equifax exposed the personal information of 147 million Americans, 15 million British consumers, and 19,000 Canadians. The data included names, birth dates, addresses, driver's license numbers, and Social Security numbers. The breach occurred through an unpatched vulnerability in a web application.

Equifax paid $700 million in settlements and remediation. Systems stabilized over several weeks, but the damage to identity was permanent. A Social Security number cannot be replaced. Millions of people learned that their most sensitive data had been entrusted to an institution they did not choose and could not leave. Trust failed because the system controlling identity lacked resilience and accountability.

Impact:
- Individuals affected: 147 million in the US and over 15 million globally
- Financial cost: $700 million
- Downtime: Weeks of disruptions and manual verification
- Long-term impact: Permanent identity exposure
- Trust lesson: Trust collapses when individuals bear lifelong risk for institutional failure

SolarWinds: Trust in the Supply Chain (2020)

Chapter 2: The Fragility of Digital Trust

In 2020, attackers compromised the SolarWinds Orion build process and inserted malicious code into trusted software updates. About 18,000 customers downloaded the infected update. The list included the US Treasury, DHS, the Pentagon, Microsoft, Cisco, FireEye, and multiple Fortune 500 companies.

The backdoor allowed attackers to access email systems and cloud accounts for months. The total remediation cost across public and private sectors reached billions of dollars. Many organizations required nine to twelve months to rebuild and verify their environments. The attack revealed that inherited trust is fragile. Compromising one vendor and thousands of organizations will inherit that compromise immediately.

Impact:
- Customers affected: 18,000
- High-value targets breached: At least 100
- Financial cost: Billions across sectors
- Downtime: Months of forensic analysis and rebuilds
- Trust lesson: Trusted updates can distribute compromise at a global scale

These incidents reveal a shared pattern. Trust fails long before encryption does. Each event began with a slight weakness in identity, verification, or process. Once that weakness spread across networks, users lost confidence in the systems they depended on. The breach did not stay local. It traveled through certificates, signatures, vendors, and institutions. The pattern shows that digital trust is fragile, not because the math is weak, but because the structures around it are unprepared for failure.

Blockchain and the Fragility of Mathematical Identity

Blockchain is a shared digital ledger. It records transactions in blocks that link together to form a chain. Each block contains a timestamp, a record of activity, and a reference to the previous block. Many computers store the same ledger. This makes the record hard to change and easy to verify. No single person controls it. The network reaches an agreement by checking the same data.

People trust blockchain because it seems secure. A transaction cannot be added unless it is verified. Every transfer must include a digital signature from the wallet owner. This signature proves ownership and authorizes the transfer of funds. The network checks the signature, confirms the transfer, and updates the ledger.

Blockchain identity relies on a pair of cryptographic keys. The public key, often represented as a wallet address, is shared openly so others can send funds. The private key is secret. It is the mathematical proof of ownership. When a transaction is created, the private key generates a digital signature. The network uses the corresponding public key to verify that signature. If the signature matches, the network assumes the owner authorized the action.

This system works today because the math behind the signature is difficult to reverse. A public wallet address does not reveal the private key that controls it. The inherent difficulty of calculating that private key is the source of trust.

Quantum computing threatens that difficulty. Bitcoin and many other blockchains use algorithms that a future quantum machine could break. A quantum system running Shor's algorithm could calculate private keys from public ones. Once that happens, any exposed wallet becomes a target.

Using quantum algorithms, it derives the private key from the public information recorded on the blockchain. No intrusion is required. No malware is installed. The math simply gives way.

Once the private key is known, control is absolute. The attacker signs a transaction using the legitimate credentials. The funds move instantly. The network validates the transfer because it meets every rule of the system. From the blockchain's perspective, nothing unusual has occurred.

There is no alarm. There is no reversal. There is no dispute process. The ledger records the transaction as final.

Researchers warn that some attackers may already be saving blockchain data today. They wait for the moment quantum machines can unlock it. At that point, years of history could become vulnerable all at once.

The weakness is not decentralization. Many nodes cannot protect a system if the signature math fails. When the math breaks, the network approves forged transactions with full confidence and authority.

Blockchain remains stable only if its cryptographic foundation remains stable. That foundation must be replaced with post-quantum cryptography before quantum systems threaten blockchain security.

The Ripple Effect of Identity Collapse

When digital identity fails, everyday life falters. The collapse does not begin with governments or corporations. It starts at the edge of personal convenience and expectation.

Online Banking: Imagine checking your bank account late at night. The balance looks wrong. Transfers you do not recognize appear complete and authorized. The system shows no errors. The transactions cleared because the attacker forged your identity using credentials that looked valid.

The loss is not limited to money. It is the moment you realize the system cannot reliably distinguish you from an impostor. From that point forward, every balance check carries doubt. Every alert triggers suspicion. The account still exists, but confidence in it does not.

Social Media: A video appears from a friend claiming an emergency. The details feel slightly off, but the voice sounds right. The face looks familiar. The urgency feels real. You respond before questioning it.

Days later, you learn the video was synthetic. No account was hacked. No password was stolen. A machine learned how to imitate someone you trust. The damage is not confined to a single incident. It lingers. Every future message from that friend now carries a hint of hesitation. Familiarity no longer guarantees authenticity.

Online Shopping: You buy a product from what appears to be a familiar retailer. The website displays a valid certificate. The checkout process looks normal. Nothing raises suspicion. Yet the domain was subtly spoofed. The certificate was misissued. The payment goes through.

Thousands of people experience the same outcome. They trusted a digital symbol and assumed the vendor ecosystem would protect them. The breach does not feel like theft at first. It feels like confusion. Only later does it become clear that trust in the process was misplaced, and that visual reassurance was never proof of legitimacy.

Encrypted Communication: Imagine sending a confidential message through a secure app. You see the familiar icon. The app reports end-to-end encryption. Unknown to you, an attacker has inserted themselves into the chain, using a forged certificate to impersonate a server. Your private words travel intact to the intended recipient, but are also intercepted and decrypted by bad actors listening in.

Government Services: A government portal displays a valid security icon. The connection is encrypted. The certificate appears legitimate. It could be a tax filing system, a Social Security portal, a civilian or military pay portal, an immigration application site, or another essential public service. Yet the certificate authority that issued the credential was compromised upstream. The site looks official because, technically, it is trusted.

Citizens submit tax records, Social Security numbers, benefit elections, immigration documents, and financial details through channels they are told to rely on. Identity theft spreads through systems that carry legal authority and public expectation of protection. The attack does not exploit the visible interface. It exploits trust in the identity infrastructure behind it.

From the citizen's perspective, nothing signals danger. The symbols are correct. The process feels routine. The portal behaves exactly as expected. The harm occurs because trust was misplaced, not because rules were broken. When trust in government identity systems fails, there is no alternative path. The very systems designed to protect citizens become the mechanisms through which harm is delivered.

IoT and Home Security: Your home security system trusts firmware updates signed by the manufacturer. So do smart locks, cameras, thermostats, voice assistants like Alexa, and other connected devices that manage daily life. A forged signing key allows an attacker to deliver malicious updates that appear legitimate.

Doors lock and unlock without warning. Cameras reboot or go dark. Alarms trigger with no intrusion. Voice assistants respond to commands no one gave. Smart appliances place orders that were never requested. Heating and lighting systems behave unpredictably.

From the system's perspective, nothing is wrong. The updates are signed. The commands are authenticated. The logs appear clean. The devices obey exactly what they are told.

The Personal Shock: When trust breaks, inconvenience turns into anxiety. Most people never think about certificate chains or cryptographic keys. They think about the moment a familiar system no longer recognizes them. A login fails without explanation. A benefits account locks unexpectedly. A transaction is flagged for reasons no one can clarify.

The shock does not come from a technical failure. It comes from uncertainty. The system that once confirmed identity now questions it. The individual is left proving who they are to machines that no longer agree. This is the true cost of collapse. When trust can no longer vouch for people, confidence gives way to doubt, and everyday life becomes fragile.

Why these Cracks Matter

Digital trust is not an abstraction. It is the quiet belief that the systems around us work as expected. It is the confidence that messages reach the intended recipient, that credentials belong to the correct person, and that the information guiding daily life has not been altered. When that confidence breaks, routine becomes uncertainty.

Quantum computing and artificial intelligence will not introduce new cracks. They will widen the ones already visible. Weak identity management becomes a liability. Outdated certificates become points of entry. Complacent processes become attack surfaces. Trust fails in small steps, then all at once.

These cracks matter because trust collapses from the inside. It collapses when the signals that once anchored confidence become unreliable. It collapses when verification becomes a question rather than an assumption. Once that happens, institutions lose their most valuable asset: the belief that their systems behave correctly.

Reflection

The fragility today stems from ordinary systems under extreme pressure, causing identity, authenticity, and integrity to falter. It shows digital trust relies on routines not built for the current speed and scale.

The Zero Epoch begins with the recognition that trust is already brittle. The vulnerabilities are not speculative. They are evident in expired certificates, misconfigured identity systems, insecure supply chains, and outdated cryptographic tools. These are not failures of technology. They are failures of preparation.

The next chapter examines the force pushing this brittleness toward collapse. When the math behind our locks no longer holds, the question becomes simple. What happens when the foundation of secrecy fails? The answer determines whether digital trust bends or breaks as the world enters a computational era it is not ready to face.

Chapter 3: When Math Stops Working

When the Impossible Becomes Practical

Digital trust is built on a quiet assumption. Some problems are so complex that no computer can solve them in any practical amount of time. The difficulty of these problems forms the backbone of modern cryptography. It protects your bank account, your medical history, your identity, and the classified communications of entire governments. Encryption works because certain calculations are designed to resist brute force, even with centuries of effort.

Quantum computing challenges that assumption at its core. It introduces algorithms and hardware that do not follow the familiar limits of classical machines. It asks a simple question with profound consequences.

Chapter 3: When Math Stops Working

What happens when the math we rely on no longer behaves the way we expect? What happens when the problems once considered impossible become merely inconvenient?

If those problems become solvable, even slowly, the effect is immediate. Encryption no longer guarantees privacy. Signatures no longer guarantee identity. Confidentiality becomes uncertain across finance, healthcare, defense, and every industry built on secure communication. The systems of the last fifty years were designed with the belief that the hardness of specific math problems would never weaken. Quantum computing removes that comfort.

The Zero Epoch is not only about the collapse of perception. It is about the possibility that the locks protecting the digital world may fail within a single generation. It introduces a future where the tools designed to guarantee secrecy can no longer be trusted. Once the impossible becomes practical, the entire architecture of digital life must be reconsidered. The world will face not only new threats, but a new definition of what it means for information to be secure.

The Unseen Threat

Quantum computing once sat in research labs and physics departments. It was a scientific curiosity, interesting but remote. That has changed. Investment surged. Competition intensified. What was theoretical is now an engineering challenge. The discussion no longer asks if quantum power will arrive. It asks how soon.

The threat hides in the math. Public-key systems like RSA and elliptic-curve cryptography rely on complex problems. Factoring large numbers and solving discrete logarithms. Classical machines cannot crack these problems within any practical lifetime. That difficulty created confidence. Quantum computing rewrites that difficulty entirely.

The unsettling truth is that the field moved from speculative to strategic. Governments now treat quantum capability as a national priority. Corporations race to prepare. Cryptographers argue about timelines. The future remains uncertain, but the consequences are not.

Breaking the Locks

Modern encryption feels indestructible because the numbers involved are enormous. A 2,048-bit RSA key is too large for classical brute force. Even a supercomputer would need billions of years to factor it.

Quantum computing changes the rules. In 1994, Peter Shor discovered an algorithm that can break RSA and elliptic-curve cryptography once scalable quantum hardware becomes available. The algorithm shortcuts the very hardness that makes these systems safe.

Grover's algorithm offers a different problem. It accelerates brute-force attacks on symmetric encryption by reducing the effective key strength. A 128-bit key behaves like a 64-bit key under quantum pressure. Security standards that once seemed stable now require recalibration.

The concept is straightforward. A safeguard designed to survive centuries may open in minutes once quantum capability matures. The world built its digital infrastructure with the confidence that these safeguards would remain unbreakable. Quantum computing challenges that belie the structure of those protections and reveal the paths that can compromise them.

Harvest Now, Decrypt Later

Chapter 3: When Math Stops Working

The quantum threat is not waiting for hardware. It began years ago.

Intelligence services and adversaries collect encrypted data today, planning to decrypt it later when quantum systems mature. This strategy is known as harvest now, decrypt later. It turns time into a weapon. Attackers do not need the ability to break encryption today. They only need the confidence that someone will break it tomorrow.

Encrypted emails sent in 2024 may be readable in 2034. Trade secrets stored today may become open files. Classified diplomatic cables protected under RSA may become readable history for hostile governments. Medical records, financial archives, and research data will not lose their sensitivity. The personal details of a child born this year could become accessible before that child reaches adulthood. A decade of corporate plans, legal negotiations, and private conversations could become transparent to anyone who captured the ciphertext while the world assumed it was safe.

This is the quiet danger. The breach does not occur when the encryption breaks. The breach occurs when the data is stolen. Decryption is delayed exposure. By the time the world realizes what has been unlocked, it will be reacting to decisions made years earlier. The timeline of risk stretches backward. Organizations inherit the consequences of their past security decisions.

This is why securing systems today is essential. It prevents adversaries from building archives of encrypted material that will become legible the moment quantum capability matures. Once data is harvested, the damage is permanent. You cannot revoke what has already been copied. You cannot reclaim a conversation intercepted a decade earlier. You cannot protect a strategy memo that was secured with algorithms already marked for retirement.

Quantum readiness is not preparation for a future threat. It is a defense against a present one. The harvest has already begun. The interception occurs now. The classification of data as harmless or low risk becomes meaningless once decryption becomes trivial. Sensitive material remains sensitive across decades, not months.

The locks hold today, but new keys are already being cut.

Lessons From DES

History offers a clear warning about what happens when trust in old protections outlives reality. The Data Encryption Standard, known as DES, once guarded the digital world. Banks used it to protect transactions. Governments relied on it to secure sensitive data. Industries across the globe treated it as a reliable lock on valuable information. When DES was introduced in the 1970s, it felt strong enough. The key was long for its time. Breaking it would have required so much computing power that the risk felt theoretical. Decision makers believed they had decades of safety. That belief slowly hardened into habit.

By the late 1990s, that assumption collapsed. Computing power had advanced far faster than expected. In 1998, the Electronic Frontier Foundation built a specialized machine that cracked DES in about two days. What once looked unbreakable fell quickly and predictably. No clever trick was required. The math did not fail. Time did.

The real danger came after DES was known to be broken. Many organizations did not move on. Systems were expensive to replace. Vendors delayed upgrades. Leaders assumed attackers would not bother with outdated targets. As a result, DES remained in use long after it should have been retired. Sensitive data stayed protected by locks everyone knew could be opened.

The lack of public incidents linked to DES did not mean the risk was low. It meant the exposure was largely invisible. Many systems protected by DES were internal business systems or government networks, not public-facing services. When those systems were accessed, there was no requirement or incentive to tell anyone.

Organizations also avoided admitting they were relying on encryption that experts had already declared unsafe. Acknowledging that reality would have raised legal, financial, and reputational questions they were not prepared to answer. Silence felt safer than transparency.

Attackers had no reason to announce success. Breaking DES did not require damaging systems or triggering alarms. It allowed quiet access to information that could be copied, studied, or exploited without leaving obvious traces. For an attacker, secrecy was the advantage.

At the time, disclosure laws were minimal. There were no modern breach notification requirements, no mandatory reporting timelines, and little public expectation of transparency. As a result, compromise could exist for years without ever becoming a headline.

This is why DES faded out quietly rather than catastrophically. Its failure was real, but it unfolded in silence.

This is the most important lesson. The greatest risk was not that DES failed. It is that people waited too long to accept that it had failed. Comfort replaced urgency.

Familiar systems felt safer than new ones. That delay extended exposure year after year.

DES teaches a simple truth. Security does not fail all at once. It fails when the change is obvious, but the action is postponed. When the math changes, waiting does not preserve safety. It quietly removes it.

The Rise of AES

The Advanced Encryption Standard, AES, emerged after the failure of DES made continued reliance impossible. DES did not collapse overnight. It eroded as computing power improved and shortcuts accumulated. Replacing it required years of analysis, standardization, vendor adoption, and hardware refresh. Nearly a decade passed before AES became the default across governments and industry. That timeline matters because it shows how slowly cryptography changes, even when failure is widely accepted.

AES represented a shift toward stronger mathematical foundations and wider security margins. It removed known weaknesses and resisted the techniques that broke its predecessor. That strength still matters today. Quantum computing threatens many classical public-key algorithms outright, but symmetric encryption behaves differently. Quantum attacks reduce AES security through brute force acceleration, not structural collapse. AES is weakened, not broken.

AES-256 is a widely used encryption standard that protects data using a huge shared secret. The "256" indicates key size, which determines how hard the encryption is to break. Because of that distinction, AES-256 remains a viable choice for long-term data protection. Its larger key size preserves a meaningful security margin even under quantum pressure. This resilience is not accidental. It reflects deliberate design choices that anticipated future advances in computation. AES reminds us that quantum capability does not erase every defense. It changes the cost equation and forces stronger parameters.

The lesson carries forward. Transitioning to post-quantum cryptography will not be a simple replacement. Unlike the DES-to-AES shift, this migration affects everything. Hardware security modules, embedded devices, operating systems, protocols, applications, certificates, and policy frameworks must all change together. Interoperability must be preserved while algorithms rotate beneath live systems.

This will be the largest cryptographic migration ever attempted. It will take longer than previous transitions, cost more, and fail unevenly where preparation is weak. AES shows that survival is possible when margins are built early. The question facing organizations now is whether they will treat post-quantum migration as an emergency later or as disciplined engineering work that begins while defenses still hold.

The Rise and Fall of RSA

RSA, named for its creators Rivest, Shamir, and Adleman, became the foundation of digital trust because it solved a critical problem at scale. It allowed unknown parties to authenticate each other, exchange secrets, and verify authority across open networks. Secure web traffic, email, software updates, VPNs, and identity systems all grew on top of RSA. It did not protect bulk data. It established belief. If RSA validated an action, systems accepted it as legitimate.

That belief still holds today. RSA remains widely deployed across global infrastructure. Transport Layer Security or TLS certificates, enterprise authentication platforms, code signing pipelines, and government systems continue to rely on RSA because it works against classical threats. It is standardized, interoperable, and deeply embedded in hardware, firmware, and protocols. Replacing it is not a configuration change. It is a coordinated reconstruction of an entire enterprise.

RSA's strength rests on a single mathematical assumption. Factoring very large numbers is infeasible for classical computers. Quantum computing breaks that assumption entirely. Shor's algorithm does not weaken RSA gradually. It renders the problem tractable. When that threshold is crossed, RSA does not degrade. It collapses. Private keys can be derived. Signatures can be forged. Identity can be impersonated retroactively.

This failure is more dangerous than data exposure. AES may continue protecting encrypted content, but RSA governs who is trusted to speak, approve, update, and command. When RSA fails, systems lose the ability to prove origin and authority. Traffic may remain encrypted, yet nothing can reliably confirm who sent it or whether it should be trusted.

This is why RSA represents a warning shot for the Zero Epoch. Its continued use assumes a future that quantum computing invalidates. Once sufficient quantum computing capability exists, systems that rely on RSA shift from secure to vulnerable without a transition period.

RSA's legacy is not failure. It carried global digital trust for decades. Its collapse marks the boundary of that era. Replacing it will require new algorithms, new certificates, new protocols, and new operational discipline across every sector that relies on digital identity. This migration will be broader and more disruptive than any cryptographic transition before it.

RSA made the digital world scalable. Its decline defines the urgency of preparing for what comes next.

Passport and Locked Suitcase

Before comparing RSA and AES, it helps to understand the difference between asymmetric and symmetric cryptography. Asymmetric systems use two related keys, one public and one private. The public key can be shared openly. The private key must remain secret. This design allows strangers to establish identity and trust across open networks. Symmetric systems use a single shared secret. Both parties must already possess the same key before communication begins. This approach is faster and well-suited to protecting large amounts of data, but it depends on establishing trust beforehand.

This distinction matters because quantum computing does not affect these systems in the same way.

Asymmetric cryptography is a passport. It proves who you are to someone who has never met you. Border agents do not share a secret with you in advance. They rely on a system that verifies identity from public information and hidden proofs embedded within the passport. If passports can be forged perfectly, borders do not weaken gradually. They stop working.

Symmetric cryptography is a locked suitcase. If you're using a TSA-approved lock, the TSA has matching (symmetric) keys that can unlock your suitcase if a search is necessary. Once you are allowed through the border, the lock protects what you carry. If thieves have better tools to pick the lock, you choose a stronger one. The suitcase does not define who you are. It protects what belongs to you.

Quantum computing forges the passport before it breaks the lock.

Asymmetric Collapse, Symmetric Endurance

RSA and AES solve different problems, fail in different ways, and behave very differently under quantum pressure. Comparing them clarifies why the Zero Epoch is asymmetric rather than absolute.

RSA is an asymmetric algorithm. It is used for identity, key exchange, and trust establishment. It answers questions like who are you, can I trust this connection, and can we agree on a secret. Its security depends on the practical difficulty of factoring huge numbers. Classical computers struggle with that problem. Quantum computers do not. Shor's algorithm turns RSA from hard to trivial once sufficient quantum capability exists. When RSA fails, it fails completely. Keys can be derived, signatures forged, and trust mechanisms collapse.

AES is a symmetric algorithm. It is used to protect data once trust has already been established. It encrypts bulk information at rest and in transit. Its security depends on brute-force resistance, not on a single mathematical shortcut. Quantum computers weaken AES through Grover's algorithm, effectively halving its key strength, but they do not break it outright. AES degrades gracefully because larger keys restore this margin.

The practical impact is decisive. RSA is a single point of failure for digital trust. When it breaks, identity, authentication, certificates, and secure negotiation fail together. AES is a durability control. When stressed, it demands stronger parameters rather than immediate replacement. That is why AES-256 remains acceptable while RSA is not.

This distinction explains why quantum disruption feels uneven. Data encrypted with AES today may remain protected if key sizes are sufficient. Systems that rely on RSA for trust and identity become vulnerable the moment quantum decryption is practical. One algorithm collapses trust. The other strains but survives.

In short, RSA protects the agreement. AES protects content. Quantum computing destroys agreement first. When agreement fails, everything built on top of it becomes suspect, even if the data itself remains encrypted.

Defining the Zero Epoch Boundary

The Zero Epoch marks the moment when asymmetric cryptography no longer provides assurance, even though systems continue to operate. Before this point, RSA-based trust holds under accepted computational limits. After it, that assumption fails completely, regardless of uptime, performance, or apparent correctness.

This transition is absolute at the cryptographic level. Either RSA can be trusted, or it cannot. What unfolds slowly is not the failure itself, but society's awareness of it. Systems keep running. Interfaces behave normally. What disappears is proof.

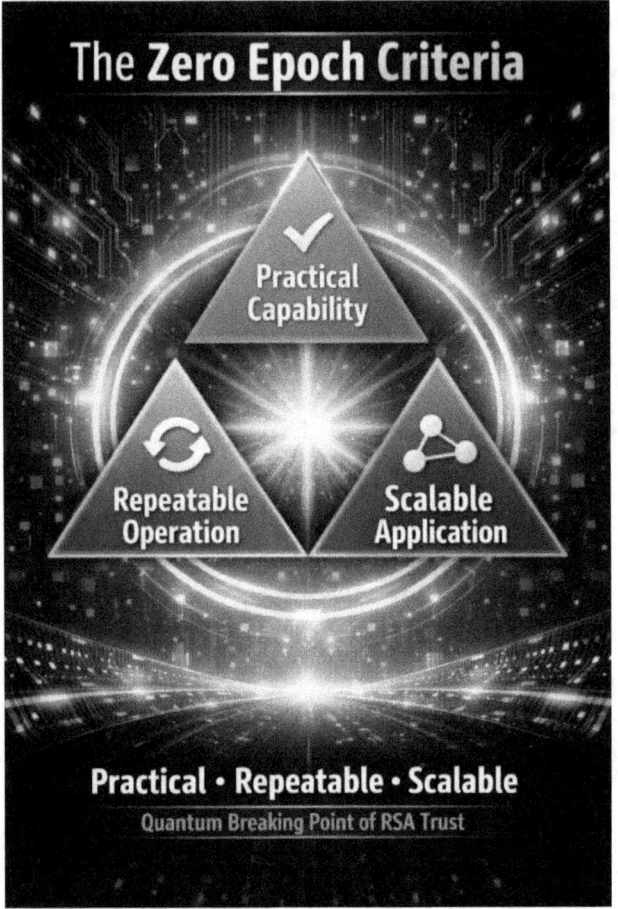

The Zero Epoch is not triggered by theoretical progress or isolated demonstrations. It is reached only when three conditions are met.

- **Practical capability.** Breaking RSA must occur within timeframes that matter to live environments. If key recovery takes years, operators can dismiss it as theoretical noise. When the same task fits inside real attack windows measured in hours or days, trust assumptions fail immediately. At that point, asymmetric security no longer protects ongoing operations.

- **Repeatable operation.** The capability must produce the same outcome across comparable quantum systems. A single laboratory success proves the possibility, not the impact. Assurance ends only when independent platforms can reproduce the result with predictable reliability. Trust collapses when failure becomes routine rather than remarkable.

- **Scalable application.** The capability must apply to many keys, not isolated demonstrations. Breaking one certificate does not alter the system. Breaking thousands turns public key infrastructure into an attack surface. At scale, asymmetric cryptography no longer limits adversaries. It feeds them with limitless opportunities.

When these conditions are satisfied, asymmetric cryptography no longer anchors trust. It becomes a liability. Public keys serve as inputs to key recovery. Digital signatures can be reproduced without the original signer. Certificate validation confirms structural correctness, not authenticity.

Nothing appears to go wrong when this happens. Encryption continues without issue, authentication still succeeds, and dashboards stay green. Systems operate as designed, but their results no longer verify their claims. Trust is still asserted, but it's just a veil of false security.

Existing language describes gradual cryptographic weakening. It does not describe the moment when trust becomes unprovable, even though everything still works. The Zero Epoch exists to name that boundary as a formal research construct, to address this gap, and to provide a unifying research lens linking cryptographic thresholds to systemic risk.

The Timeline Nobody Knows

Experts disagree on when quantum machines will break RSA. Some predict decades. Others believe it may arrive far sooner because of investment from China, the United States, Europe, and private industry. Each research breakthrough shortens the horizon. Each financial push increases the pace. The uncertainty becomes the risk.

When people believe they have time, they wait. They defer upgrades. They postpone migrations. They treat preparation as optional. If the breakthrough arrives early, that waiting becomes a failure. The risk does not come from the date itself. It comes from the assumption that the date is distant.

Quantum machines today are small and noisy. They operate at temperatures colder than space. They require isolation from vibration and electromagnetic interference. Qubits are the basic units of quantum computers. Unlike classical bits, which are either on or off, qubits can exist in multiple states at the same time. This property allows quantum machines to explore many possibilities at once. Decoherence occurs when a qubit loses this delicate state due to heat, vibration, or electrical noise. When decoherence happens, the quantum advantage disappears and the system behaves like an ordinary computer. Keeping qubits stable long enough to perform useful work remains one of the hardest problems in quantum computing. Materials science, fabrication, and error correction all pose challenges. Yet progress continues at a steady pace. New qubit designs emerge. Error rates fall. Coherence times improve. Each year pushes the boundary closer.

Waiting for a perfect machine misses the point. The threat does not require perfection. It requires a capability above a certain threshold. Once hardware

Chapter 3: When Math Stops Working

reaches that threshold, the locks that protect modern digital life collapse at once.

History shows that transformative technology crawls slowly and then sprints. Early electronic computers were fragile and limited, then became essential within a generation. Smartphones were niche tools, then reshaped global communication within a decade. Machine learning delivered modest advances for years, then surged as computing power became sufficient to process vast amounts of data at scale. Once the pieces fit, acceleration became unavoidable.

Quantum computing sits at the edge of the same pattern. The world will experience a long period of apparent stagnation, followed by a rapid shift that outpaces planning cycles. The challenge is that no one knows when the sprint begins.

Y2K: The Countdown That Humanity Beat

The world faced a digital deadline once before. Year 2000 or Y2K was not a false alarm. It was a global intervention. Computers became mainstream between the late 1970s and the mid-1980s. In those days, computers stored years with two digits instead of four for efficiency because computing power was limited, slow, and costly.

An alarm was raised, and the world took Y2K seriously in 1996 when governments, large corporations, and the media recognized Y2K as a systemic risk rather than a technical curiosity. The problem was clear. When systems rolled from 99 to 00, many would treat the year 2000 as 1900. Banking platforms, flight control software, medical devices, and power grid schedulers all depended on accurate time. A failure in any of them could have triggered cascading outages.

The threat created a clear timetable: midnight on December 31, 1999. Organizations understood the deadline and treated it with a sense of urgency. Governments funded large remediation programs. Companies reviewed millions of lines of code. Engineers documented every system that stored a date and tested every patch. The United States spent an estimated $100 billion. Worldwide spending passed $300 billion.

The preparation worked. When the moment arrived, banks opened. Planes flew. Hospitals operated normally. The world witnessed calm because the work began early and continued without pause. Y2K did not vanish because it was small. It vanished because millions of people acted before the problem matured.

The comparison matters. The Zero Epoch has no countdown. There is no shared deadline. The threat is invisible and resides quietly within systems that rely on cryptographic strength. Y2K showed the power of coordinated preparation with a fixed target. The Zero Epoch shows the difficulty of preparing when the target is unknown.

Disruption is avoided only when leaders act before the failure, not after. Quantum development moves without a calendar. Governments cannot mark a date on the wall. Organizations cannot plan for this midnight event. This uncertainty becomes a risk because delays are easier when the deadline is unclear.

Accelerators: Money, Strategy, and Competition

Quantum development is no longer a scientific curiosity. It is a geopolitical race. Nations treat quantum research as a strategic asset because the winner gains an advantage that reshapes intelligence, cyber defense, and global influence.

Chapter 3: When Math Stops Working

The United States invests through national initiatives, research labs, and private sector partnerships. Federal programs coordinate academic work with commercial hardware development. Venture capital fuels startups that build quantum hardware, simulation platforms, and post-quantum security tools. The strategy is clear. Build the strongest ecosystem and accelerate progress through competition.

China embeds quantum advancement directly into national planning. It funds research centers, satellite experiments, secure communication networks, and military applications. State-backed programs push rapid deployment. Success is measured in capability, not publication. Quantum technology is treated as part of national power and their progress may not be made publicly.

Europe coordinates quantum science through large-scale programs that unify research institutions across member states. Their focus includes hardware, materials science, communication infrastructure, and standards that define global interoperability. Europe aims to shape the rules that govern quantum technology as much as the technology itself.

Other regions are moving as well. Japan invests in quantum sensors and industrial applications. Australia advances quantum hardware and defense integration. India accelerates quantum communication research through national labs.

Each region sees quantum supremacy as a lever of influence. The stakes exceed scientific achievement. Whoever arrives first gains the power to unlock the world's encrypted archives. Intelligence reports, private transactions, diplomatic cables, intellectual property, and classified communications become transparent to the side that has a quantum advantage.

The prize is not speed. It is access to information, knowledge, and secrets. Information becomes leverage. Leverage shapes negotiation, deterrence, and strategic posture. Quantum capability becomes a form of power that does not fire a shot but shifts the balance of global supremacy.

The race is already underway. The question is not who participates. The question is who prepares before the moment of advantage becomes irreversible.

The Paradox of Uncertainty

The greatest danger is not the quantum machine that breaks encryption. There is uncertainty about when that machine will appear.

Preparation takes years. Migration requires planning. Yet leaders hesitate, hoping for clarity to justify reallocating limited resources. That hesitation becomes the risk. The move from DES to AES took years. The transition to longer RSA keys took decades. The subsequent migration will be larger, broader, and more expensive.

Even if the first practical quantum computer arrives twenty years from now, the threat exists today. Every encrypted file stolen today becomes readable later. The future is already compromised. It is simply not decrypted yet.

Reflection

The Zero Epoch teaches a simple lesson. Defenders cannot plan on timelines. They must plan on the certainty of impact. Waiting for confirmation is not strategy. It is surrender. A quantum system capable of breaking today's encryption is inevitable. The collapse of current cryptographic assumptions is guaranteed. The exact year is irrelevant.

Chapter 3: When Math Stops Working

Y2K had a date. Quantum disruption does not. Y2K had a bug confined to a specific flaw. Quantum disruption affects an entire ecosystem of systems that rely on public-key cryptography. Y2K had a coordinated global response built around a known deadline. Quantum disruption demands a decade of preparation with no countdown and no shared moment of urgency.

The Zero Epoch marks the moment the math stops working. It is the point at which quantum capability breaks the locks that protect identity, finance, healthcare, government, and national security. Human attackers exploit mistakes. Quantum machines remove the need for mistakes. They eliminate the friction that once acted as a barrier to secrecy.

If Chapter 2 revealed the early cracks in digital trust, Chapter 3 revealed the force that widens them. The next threat is artificial intelligence, which targets perception instead of secrecy. Quantum breaks the locks. AI breaks reality. Together, they create a world where neither proof nor perception can be accepted without scrutiny. Chapter 4 examines this shift. It explores the moment when machines do not simply compute. They learn to impersonate, influence, and deceive.

Chapter 4: When Machines Learn to Lie

The Age When Perception Becomes a Weapon

The first threat to digital trust comes from quantum power. The second comes from something far more human. Artificial intelligence does not break encryption. It breaks confidence. It forges faces, voices, and documents with such precision that the senses become unreliable. Where quantum computing attacks our locks, AI attacks our perception.

Humanity has always faced deception. What changes in the Zero Epoch are the speed, scale, and accuracy of the lie. AI industrializes manipulation. It learns from every attempt. It adapts without fatigue. The result is a world where evidence competes with simulation and certainty erodes in the noise.

This chapter explores that erosion. It shows how machines learned to mimic humanity and how quickly they turned that skill against the systems that depend on truth.

The Nature of Perfect Imitation

AI does not deceive people by outsmarting them. It deceives them by imitating them. It studies patterns in voice, posture, writing, and expression. It learns how people sound when they are confident, tired, angry, or afraid. It observes how individuals structure sentences, pause between thoughts, and signal emotion. Then it reproduces those patterns with precision.

This is the core problem. Humans trust what feels familiar. A familiar face. A familiar voice. A familiar signature. Familiarity has guided survival for thousands of years. AI exploits that instinct. It creates illusions that bypass critical thinking and target emotion. The deception works not because it is clever, but because it feels comfortable.

The danger is not only the lie. It is the erosion of belief that follows. When any video can be forged, people question all videos. When any recording can be fabricated, people doubt every recording. When every document can be synthesized, authenticity becomes uncertain. Once that uncertainty becomes constant, trust collapses because no signal feels definitive.

AI changes the environment of perception. It replaces the instinctive cues that once established confidence with simulations that are indistinguishable from reality. Humans were not built to navigate a world where imitation is perfect, and authenticity is optional.

Deepfakes and Synthetic Identities

Deepfakes began as strange internet curiosities. They were viewed as experiments rather than threats. Today, they are dangerous tools. AI can generate faces that never existed, clone a voice from a few seconds of audio, and animate a figure in real time. The forgeries no longer look artificial. They look familiar.

Every synthetic face is a mask. Every cloned voice is a performance. The illusion works because it imitates human rhythm. It captures the slight hesitations, the imperfect breathing, the natural breaks in speech. Imperfection makes the forgery more convincing, not less. A deepfake does not need to be flawless. It needs to feel real long enough to trigger action.

Synthetic identity changes the meaning of authenticity. Verification used to depend on photographs, recordings, and signatures. These artifacts once carried authority because producing them required presence and effort. In the Zero Epoch, they become uncertain. A verified image becomes a suggestion. A recorded message becomes a possibility. Evidence transforms from a foundation into a variable.

The threat does not come from a single forgery. It comes from a world where every piece of evidence must defend itself before it can be trusted.

The Human Challenge

The human mind evolved to make fast judgments. We trust our senses because they helped our ancestors survive. That instinct becomes a liability in a world shaped by AI. Our perception becomes the weakness that AI exploits with precision.

The emotional toll is deeper than any technical breach. When people can no longer trust what they see or hear, they begin to question everything. That constant doubt becomes exhausting. Fatigue leads to two outcomes. Some people blindly believe everything because skepticism requires too much effort. Others challenge nothing or contribute to the discourse because the risk of being wrong feels too high. Both conditions are dangerous because they destabilize public life.

Even trained professionals feel the pressure. Analysts become cautious when evidence appears valid but arrives in unfamiliar ways. Leaders hesitate when a voice sounds correct, yet could be fabricated with ordinary tools. Citizens read a statement and wonder if the speaker ever said the words. The shared sense of reality weakens. The world loses coherence.

AI does not simply impersonate people. It impersonates authority. It impersonates urgency. It impersonates fear. These emotional cues influence decisions faster than logic and faster than verification. They shape behavior while people are still deciding whether to pause and think.

The human challenge is not only technical. It is psychological. It forces society to confront the limits of intuition in an environment where intuition can no longer be trusted.

Adversarial AI and Machine Manipulation

AI is not only a manipulator. It is also vulnerable to manipulation. Machine learning models can be misled by subtle distortions invisible to human eyes. A single altered pixel can convert a correct classification into a dangerous mistake. Is it inconceivable that a sticker on a stop sign can make a self-driving car accelerate?

This is the new attack surface. It is no longer defined only by servers or firewalls. It is characterized by perception, interpretation, and classification. When an artificial intelligence system misreads reality, the consequences extend beyond data and into the physical world.

An image is misclassified. A command is misinterpreted. A pattern is detected where none exists, or missed where it matters most. These failures expose a simple truth. An AI system does not understand the world the way a human does. It does not see, hear, or reason about reality. It processes data patterns.

When an AI analyzes an image, it is not recognizing an object. It is matching pixels to statistical patterns it learned during training. When it interprets text or speech, it is not grasping meaning or intent. It predicts the most likely output based on prior examples.

Because of this:
- AI has no awareness of context outside its data.
- AI has no intuition or judgment.
- AI cannot tell truth from deception on its own.
- AI cannot recognize when it is being misled.

It produces answers that look confident because confidence is a pattern it has learned to display. That confidence does not imply correctness.

This is why the phrase matters. When AI does not perceive reality, it can be manipulated by inputs that appear normal but are designed to trigger mistakes. In the Zero Epoch, those mistakes can move money, reroute systems, or influence people before anyone realizes the interpretation was wrong.

In short, AI does not know what is real. It only knows what looks statistically familiar. It reproduces patterns learned from prior data. That makes it powerful, but also vulnerable. Carefully crafted inputs can exploit its blind spots, turning confidence into error and automation into risk.

Scenarios and Vignettes

These incidents show how synthetic content moves faster than human judgment. They reveal how quickly trust can collapse when a machine imitates authority or fabricates urgency. None of these scenarios require technical genius. They need only a convincing illusion and a human being pressured to decide quickly.

Below are nine real events. Each one shows a different facet of AI-enabled deception. Each one marks another step toward a world where machines are taught to lie at scale.

AI-Generated Nude Photo Extortion Rings (February 2020)

In February 2020, coordinated extortion rings used AI tools to create synthetic explicit images of real women. Attackers harvested public photos from social media platforms and used image-generation software to fabricate nude content. Victims received demands for payment under threat of exposure. Law enforcement tracked activity across more than 20 countries and thousands of reported cases, with many more unreported. Financial impact varied. Many victims paid small amounts to avoid embarrassment. The emotional cost was substantial. Restoration required takedown requests, forensic reviews, and law enforcement involvement, often lasting weeks.

The incident showed how identity can be weaponized even without direct access to a victim.

Impact:
- Victims: Thousands across more than 20 countries
- Financial loss: Varied, but widespread small-payment extortion
- Downtime: Weeks of investigation and takedown coordination

- Restoration: Case-by-case deletion and legal support
- Trust lesson: AI can fabricate identity violations that feel real and immediate

Deepfake Impersonation of UAE Royals for Investment Fraud (September 2021)

In September 2021, scammers cloned the voice and face of a member of the UAE royal family to solicit false investment opportunities. Attackers produced convincing video calls and voice messages that mimicked official communications. Organizations across the region received fabricated investment requests worth millions of dollars. Some victims transferred funds before discovering the deception. Recovery required legal intervention and diplomatic coordination. Several companies paused cross-border transactions for days while verifying identities.

This incident showed that political authority can be forged as easily as a corporate executive.

Impact:
- Targets: Multiple companies and high-net-worth individuals
- Financial loss: Attempts worth millions, some confirmed losses
- Downtime: Days of suspended transactions
- Restoration: Verification through government channels
- Trust lesson: High-profile identity can be replicated with minimal evidence of fraud

Deepfake Celebrity Endorsement Scams (December 2022)

In December 2022, attackers created AI-generated videos of well-known celebrities endorsing fraudulent investment schemes. The footage used accurate facial motion, realistic lighting, and convincing vocal reproduction. Victims believed the endorsements and invested money into nonexistent platforms. Individual losses reached tens of thousands of dollars. Companies hosting the fraudulent ads faced forced takedowns and account reviews. Victims spent weeks attempting to reverse payments through banks and payment processors.

This incident highlighted that familiar public figures no longer anchor truth when their likeness can be recreated at will.

Impact:
- Victims: Thousands globally
- Financial loss: Tens of thousands per individual
- Downtime: Weeks of refund claims and reversals
- Restoration: Platform takedowns and identity verification
- Trust lesson: Celebrity recognition cannot validate authenticity

UK Energy Firm Voice Clone Fraud (March 2019, publicized in 2023)

An executive at a UK energy firm received a call that sounded identical to his CEO. The voice referenced ongoing projects. It spoke with the correct accent and tone. The executive authorized a transfer of 220,000 euros. Investigators confirmed that the voice was generated by an AI model trained on public recordings. Internal operations slowed for several days as teams reviewed earlier communications and froze pending transfers.

The incident demonstrated that mimicry no longer requires human skill. A machine can imitate authority convincingly.

Impact:
- Organization affected: 1 energy firm
- Financial loss: 220,000 euros
- Downtime: Days of internal review
- Restoration: Manual audit of communications
- Trust lesson: Voice authentication cannot stand as a security control

Slovakia Election Deepfake Audio (September 2023)

Two days before Slovakia's national election, a synthetic audio clip spread across social platforms. It appeared to show opposition leaders discussing plans to rig the vote. The recording reached tens of thousands of users within hours. Journalists debunked it the same day, but voter sentiment had already shifted. Election officials could not retract or suppress the clip. The disinformation circulated until the polls opened.

The incident demonstrated how democratic trust can be bent by a well-timed synthetic message.

Impact:
- Population exposed: Tens of thousands
- Financial loss: Not disclosed
- Downtime: No system outage, but political disruption persisted
- Restoration: Media clarification, not technical removal
- Trust lesson: Elections can be influenced faster than truth can respond

Arup Hong Kong Deepfake Conference Fraud (January 2024)

In January 2024, Arup staff joined a video call with people they believed were company directors. Every face and voice except the victim's was synthetic. Over the course of the call, the employee approved transfers totaling 25.6 million dollars. Internal operations froze for days. All pending transactions required manual verification. Audits stretched into weeks. The organization implemented new controls based on physical confirmation.

This incident showed that video presence, once considered reliable, can now be forged convincingly.

Impact:
- Financial loss: About 25.6 million dollars
- Downtime: Days of suspended financial operations
- Restoration: Internal audits and cross-team verification
- Trust lesson: Video identity cannot function as a security guarantee

AI-Generated Voice Robocalls Impersonating a Public Official (September 2024)

A political consultant used AI to clone a public figure's voice. The synthetic calls urged recipients to avoid voting in an upcoming primary. Regulators traced the operation and imposed a $6 million fine. Restoration required rapid public communications and voter education. Election administrators spent days countering misinformation.

The incident showed that synthetic speech can directly influence civic participation.

Impact:
- Population affected: Thousands of voters
- Financial penalty: 6 million dollars
- Downtime: Days of election-related confusion
- Restoration: Public announcements and voter guidance

- Trust lesson: Civic communication cannot rely on voice identity

Deepfake Used to Access Corporate VPN (January 2025)

Attackers used synthetic face and voice generation to pass a corporate video-based multi-factor authentication check. The intrusion was limited, but the audit cost ran into the millions. Security teams paused remote access systems for days. Verification procedures were rewritten.

The incident showed that identity checks based on appearance are no longer dependable.

Impact:
- Organization affected: Major US tech firm
- Financial loss: Millions in audit and recovery
- Downtime: Days of restricted remote access
- Restoration: Revised authentication and controls
- Trust lesson: Appearance-based verification is no longer secure

AI-Generated Investment Brief Scam (March 2025)

Synthetic financial briefings mimicked the tone, structure, and formatting of a major investment bank. The briefings circulated in trading groups. Market activity shifted before the hoax was flagged. Firms spent days reviewing communications and instructing analysts to disregard the forged reports.

The incident revealed how easily AI can imitate professional authority.

Impact:
- Entities affected: Multiple trading desks
- Financial loss: Millions in short-term market movement

- Downtime: Days of communication review
- Restoration: Manual verification of research reports
- Trust lesson: Expertise signals can be replicated without real expertise

The Pattern Behind the Incidents

Each event exposed a different fracture in digital trust. Some targeted what people see. Others targeted what people hear. Some manipulated market signals operate faster than human judgment. Others exploited political tension or the predictable rhythm of professional work. The methods varied. The weakness was constant. Modern systems still trust signals that machines can now imitate. Across these incidents, five themes emerge with clarity.

Identity signals fail first: Voice, face, signature, video presence, and official symbols become unreliable the moment a machine can reproduce them. The cues that once felt intuitive lose meaning. Familiarity becomes a liability. A world built on recognition begins to fracture the instant those signals can be forged at scale.

Markets move faster than verification: Synthetic content triggers earnest financial reactions before analysts or systems can respond. Trading engines react to images, headlines, and sentiment in milliseconds. By the time the truth arrives, the damage has already taken shape. The market does not wait for confirmation. It moves on perception.

Election processes and civic trust degrade: Synthetic messages distort public judgment. They target emotions, not facts. They spread faster than rebuttals and linger longer than clarifications. Even after corrections appear, doubt remains. Confidence in democratic events weakens when citizens cannot distinguish authentic speech from artificial manipulation.

Professional and scientific credibility becomes fragile: Fake reports and forged documents imitate expertise with a precision that once required human skill. They borrow the tone, structure, and references of legitimate institutions. They influence decisions before verification begins. They weaken trust in the processes that safeguard medicine, engineering, finance, and research.

Verification shifts from automatic to manual: Organizations fall back on audits, phone calls, cross-checks, and emergency reviews. They move from efficiency to certainty. They operate at human speed because machine verification can no longer guarantee truth. This shift happens after the damage is done, not before.

These patterns show the same trajectory. Trust erodes before systems fail. Confidence dissolves before security breaks. The Zero Epoch magnifies this path by accelerating the forces that exploit these weaknesses. The cracks visible today are not isolated events. They are early signs of a structure under strain.

AI does not attack technology alone. It attacks the assumptions people rely on. The belief that a familiar voice must be authentic. The assumption that a document must be legitimate if it looks official. The assumption that urgency implies truth. The assumption that a video call must be genuine because the movement appears natural.

These cases show how quickly trust collapses when attackers imitate the signals that people once treated as proof. The Zero Epoch requires that these assumptions be questioned before they become points of failure.

When Reality Becomes Optional

The danger is not the lie. The threat is the condition that follows the lie. It is a world where truth requires forensic validation. It is a world where leaders hesitate because every command might be synthetic. It is a world where citizens select only the facts that comfort them because objective evidence feels optional.

AI deception scales faster than human cognition. It spreads faster than the law. It adapts faster than policy. It reaches audiences at a speed that outpaces instinct and overwhelms judgment. People trust their senses until the day their senses betray them. By the time they realize something is wrong, the pattern is already established.

The collapse of trust will not feel like a single moment. It will feel like a slow drift. Certainty fades. Familiar cues become unreliable. Doubt becomes normal. Society begins to treat information not as a shared reference point, but as a menu of competing realities. Once this shift begins, restoring consensus becomes difficult because every claim must first defend its existence.

AI does not merely challenge truth. It changes the environment in which truth operates. It creates conditions in which reality becomes negotiable, and proof becomes a burden rather than a foundation. That is the real threat. It is not the deceptive artifact. It is the world that emerges once people accept that deception is effortless.

Reflection

Quantum computing breaks math. Artificial intelligence breaks perception. Together, they weaken the foundation of digital trust from two sides. One erodes the ability to secure information. The other erodes the ability to recognize what is real. The combination produces a world where identity, authenticity, and certainty no longer behave as society expects.

The following chapter explores whether governance can keep up with such rapid disruption and whether authority can stay credible in a world where truth is flexible and trust needs rebuilding in a context driven by machine speed. If traditional safeguards break down and reality becomes negotiable, the question arises: can institutions, designed in an era of analog technology, oversee a future dominated by systems that learn, deceive, and evolve faster than laws can be established?

Chapter 5: When Governance Cannot Keep Pace

When the Past Tries to Govern the Future

Governance was built for a world that moved slowly. Laws were written on the assumption that facts remained stable, institutions remained predictable, and threats evolved in visible patterns. That world is gone. Artificial intelligence and quantum computing move at machine speed. Governance moves at human speed. The gap between them widens every year.

The Zero Epoch exposes this mismatch with uncomfortable clarity. Technology now reshapes itself in weeks. Policy still arrives in seasons. Oversight strains to understand systems that learn, adapt, and reorganize before committees even meet. What once felt like stable ground now feels like shifting sand.

This chapter examines that gap. It shows why the rules fall behind the risks and why governance must become faster and more fluent if trust is to survive. As Chapter 3 showed through the Y2K response, governance succeeds only when action begins before the risk peaks.

The Lag of the Law

Law moves at human speed. It requires language, consensus, and process. Drafts are written. Committees debate intently. Revisions follow compromise. Votes conclude months or years later. This pace assumes that the threat environment remains stable long enough for rules to matter. Modern technology rejects that assumption entirely. An AI model can be trained, deployed, and iterated globally in days. A quantum breakthrough can propagate through research communities before regulators even know the problem has changed.

Governance responds the only way it knows how. It studies impact after the fact. It investigates failures. It issues guidance once harm is visible. Then it codifies lessons learned into rules meant to prevent recurrence. This approach works when threats are slow, localized, and predictable. It fails when the threat adapts faster than institutions can observe it. By the time regulation arrives, the system it was meant to govern has already moved on.

The gap is structural, not procedural. Governance assumes continuity. It assumes that past behavior predicts future risk. In the Zero Epoch, that assumption collapses. Technologies do not evolve linearly. They compound. They learn. They scale without regard for jurisdiction or approval. Yesterday's controls describe a world that no longer exists.

This creates a dangerous illusion of safety. Compliance appears intact while relevance erodes. Organizations faithfully follow the rules while operating under conditions the rules never anticipated. Regulators focus on enforcement while adversaries focus on acceleration. The result is not lawlessness, but misalignment. Law arrives informed, careful, and late.

In the Zero Epoch, resilience cannot depend on regulation catching up. Governance must shift from prescribing behavior to enforcing adaptability. Systems must be designed to withstand threats that law has not yet named. Policy still matters, but it can no longer be the first line of defense. When time itself becomes the vulnerability, survival favors those who prepare for change before permission exists.

The Speed Mismatch

Technology moves on its own clock. Software updates itself. Models retrain automatically. Systems learn from live data and adjust behavior without human approval. Oversight moves differently. Humans negotiate language, intent, and responsibility. That difference in speed creates permanent tension between innovation and control.

By the time a legislature proposes an AI transparency requirement, the next generation of models has already learned how to bypass inspection. By the time a quantum security standard is drafted, researchers have published architectures that weaken the assumptions behind it. Rules arrive based on a snapshot of reality that no longer exists. The target has already shifted.

Threat actors operate without these constraints. They do not wait for consensus or permission. They adopt new tools as soon as they appear. They move across borders, platforms, and jurisdictions without friction. While governments debate definitions and scope, adversaries test, refine, and deploy.

This mismatch shapes the modern threat landscape. Attackers iterate continuously. Governance deliberates carefully. The gap between those speeds is where exploitation thrives. In the Zero Epoch, risk is not defined by intent alone. It is defined by who can adapt fastest, while others are still deciding what to regulate.

The Patchwork Problem

There is no global rulebook for AI or quantum systems. Instead, governance is fragmented across national priorities and political cultures. The European Union emphasizes privacy and accountability through GDPR. The United States relies on sector-specific laws, agency guidance, and voluntary frameworks. China prioritizes state control paired with rapid deployment. India combines aggressive adoption with comparatively light regulation. Each approach reflects local values. None align cleanly with the others.

These differences create seams. Seams create opportunity. Attackers do not need to break the strongest system. They move through the weakest jurisdiction, the loosest requirement, or the slowest enforcement path. What is illegal in one region becomes ambiguous in another and invisible in a third.

A single breach can cascade across borders. Data stolen in one country becomes extortion material in another and a diplomatic issue somewhere else. Corporate leaders struggle to reconcile conflicting obligations. Engineers implement controls that satisfy one regulator while violating another. Compliance becomes a balancing act rather than a security posture.

The internet ignores borders. Adversaries exploit that reality. Governance does not share that advantage. Fragmentation turns compliance into confusion. Confusion becomes vulnerability. In the Zero Epoch, inconsistency is not a policy inconvenience. It is an attack surface.

The Compliance Mirage

Executives often equate compliance with security. The impulse is understandable. Compliance can be tracked. It can be audited. It produces charts, checklists, and reports. Security rarely feels that concrete. It deals with uncertainty, motion, and adversaries who do not announce their intentions. Passing an audit creates a sense of safety, yet that safety is fragile. It reflects what was true on the day of the audit, not what is true when the threat arrives.

Threat actors do not check audit calendars. They do not file paperwork. They do not respect regulatory boundaries. They search for weak points created by outdated rules, slow update cycles, and rigid approval processes. They exploit gaps that remain invisible to oversight teams until after the breach.

Compliance can create a false sense of accomplishment. It encourages organizations to treat security as a finish line rather than a continuous practice. It hides deeper weaknesses behind neatly formatted reports. It rewards routine rather than readiness. A compliant organization is not always a resilient one. Compliance ensures that the right boxes are checked. It does not guarantee that the right defenses exist.

Strong security requires agility. It involves awareness of new techniques, new vulnerabilities, and new attack surfaces. It requires continuous assessment, not annual reviews. It demands systems that evolve at the speed of the threat. In the Zero Epoch, resilience depends on preparation, not paperwork. The organizations that survive will be the ones that treat compliance as a baseline, not a shield.

The Ethics Vacuum

AI now participates in decisions that affect employment, credit, healthcare, legal outcomes, and public safety. These systems influence who receives treatment, who receives funding, who receives support, and who receives scrutiny. Yet the ethical frameworks that guide these decisions develop far too slowly. Oversight boards meet quarterly. The models change weekly. The gap widens every cycle.

Bias can spread inside a system faster than any reviewer can detect. A model trained on incomplete data can reinforce old inequities. A small error in a scoring algorithm can shape outcomes for millions. Once deployed, these systems learn from feedback loops that humans struggle to observe. Their influence grows even when their logic remains opaque.

Quantum computing introduces its own ethical burden. Nations and corporations may develop offensive decryption capabilities in secret. A breakthrough in one lab can compromise global confidentiality without public disclosure. Defensive capability requires cooperation, transparency, and shared standards. Offense requires only ambition and a decision to remain silent.

This imbalance rewards risk. It encourages acceleration without caution. Organizations push ahead because hesitation feels like a disadvantage. Policymakers respond after the damage appears, not before. Governance becomes reactive rather than strategic.

The ethics vacuum is not the absence of values. It is the absence of processes fast enough to apply those values. It is the realization that the systems shaping society evolve at speeds that the ethical frameworks cannot match. Until that changes, the most significant vulnerability will not be the technology. It will be our inability to guide it with the clarity and restraint that the Zero Epoch demands.

Governance at Cross Purposes

Innovation and regulation often distrust each other. Policymakers fear slowing progress. Engineers resent constraints imposed by people who do not understand the systems. Each side treats the other as an obstacle. Both sides lose sight of the real adversaries. The friction is not intentional. It is structural.

This tension creates paralysis. Policymakers wait for clarity. Engineers wait for direction. Each assumes the other will act first. Meanwhile, threat actors experiment freely. They face no delays, no hearings, and no approval cycles. They test new methods while the defenders debate terminology.

Governments have responded unevenly, each publishing principles or roadmaps that signal intent but not readiness. The table below highlights some of the major AI and quantum governance frameworks published in recent years. Together, these documents show that nations are preparing, but they also reveal how uneven and fragmented global oversight has become.

Country / Region	Strategy Documents and Publication Years
United Kingdom	National Quantum Strategy (2023); AI Regulation White Paper (2023)
Canada	National Quantum Strategy (2023); Directive on Automated Decision-Making (2020)
United States	Executive Order on Safe, Secure, and Trustworthy AI (2023); NIST AI Risk Management Framework (2023); CNSA 2.0 Post-quantum Standards (2022)
European Union	Quantum Technology Flagship Strategic Roadmap (2018); EU Artificial Intelligence Act (2024)
Singapore	Model AI Governance Framework (2019, updated 2020)
Japan	AI Strategy 2022 (2022); Quantum Technology Innovation Strategy (2020)
South Korea	National AI Strategy (2019); National Quantum Science and Technology Plan (2022)
India	National Strategy for Artificial Intelligence (2018); India Quantum Mission (2023)
China	Next Generation AI Governance Principles (2019); Ethical Guidelines

	for New Generation AI (2021); National quantum initiatives under Medium and Long Term S&T Development Strategy (updated 2021)
Thailand	Thailand AI Ethics Guidelines (2022)
Indonesia	Indonesia National AI Strategy, Stranas KA (2020)
Vietnam	National Strategy on Research, Development and Application of AI (2021)
United Arab Emirates	National Artificial Intelligence Strategy 2031 (2019); National Program for Artificial Intelligence (2019)

These documents outline aspirations, but none operate at the speed required by AI acceleration or quantum disruption. The gap between principle and enforcement grows wider each year. The systems evolve but the rules do not. The gap widens until oversight becomes symbolic. The pace of change overwhelms the pace of regulation. Without alignment, innovation accelerates while governance lags. This imbalance defines the Zero Epoch. It signals that the challenge is not only technical. It is structural. It is cultural. It is the failure to coordinate at the speed required by the world we have created.

Case Study

The AI Bill That Time Outran (United Kingdom, 2023)

In 2023, the United Kingdom introduced the Artificial Intelligence Accountability and Transparency Bill. It aimed to regulate high-risk AI systems by requiring training documentation, bias testing, and access to audits. The bill targeted the AI models that existed at the time it was drafted.

By the time Parliament debated it, the industry had already shifted. Developers were releasing continuously learning systems that did not match the bill's definitions. Concepts like "training completion" no longer applied. The models evolved long after deployment.

The law passed. Enforcement struggled. Fewer than two dozen specialists were assigned to monitor AI across multiple sectors. Within a year, the bill's language lagged behind the technology it attempted to govern.

The gap was not intentional. It was structural. The law aimed at a moving target and missed.

Why Bureaucracy Breaks

Bureaucracy needs stability. It depends on predictable change. It assumes that systems behave consistently enough to regulate in advance. Rules are written with the belief that technology will evolve in slow increments and that oversight will have time to study each shift before it takes effect.

AI and quantum systems break that assumption. They shift without warning. They generate new behaviors that no one anticipated. They produce outcomes that move faster than review cycles. They challenge regulations built for an age when systems advanced in steps rather than leaps.

These technologies do not wait for approval. They do not pause for interpretation. They revise themselves through constant iteration. A bureaucratic process that requires meetings, analysis, and multi-year rulemaking cannot keep pace with systems that adjust their logic within seconds.

This is not a failure by neglect. It is failure by design. Governance cannot revise rules as fast as AI can revise its models. It cannot update policy as quickly as quantum capabilities expand the attack surface. Oversight depends on comprehension, yet many decisions are now made by systems that provide results without offering explanations.

Every update creates a new blind spot. Every delay magnifies risk. A rule written today may not apply by the time it is published. A safeguard approved last year may be irrelevant to the behavior of current systems. The gap between policy and practice grows wider as capability accelerates.

The longer governance waits, the more complicated the catch-up becomes. What began as a manageable discrepancy becomes structural misalignment. Institutions designed for gradual change must now operate in an environment defined by acceleration. Without redesign, bureaucracy will always arrive after the damage has occurred.

The Illusion of Control

Society depends on the belief that rules shape outcomes. Institutions assume that decisions follow logic that can be reviewed, measured, and corrected. This belief holds as long as systems behave in ways humans can trace.

AI disrupts that expectation. Complex models generate results that even their creators cannot fully explain. Quantum processes collapse faster than they can be observed. Both operate in domains where cause and effect unfold in patterns that resist interpretation. A decision appears, yet the path that produced it remains hidden.

Accountability becomes difficult when outcomes emerge from layers of computation too complex to audit. An AI model denies a loan. A quantum simulator recommends a military posture. A predictive tool influences policing. Each decision may be helpful or harmful. The problem is not the outcome. The problem is the absence of a clear explanation for the outcome.

Governance loses its grip when it cannot trace how a conclusion was formed. Oversight depends on transparency, and transparency fails when the reasoning becomes inaccessible. Regulators ask for justification. Engineers offer probability maps and confidence scores rather than a cause. Leaders approve actions guided by models they cannot fully understand.

Without explanation, oversight becomes symbolic. People continue to authorize reviews and require sign-offs, but the process no longer guarantees insight. Control becomes an illusion because the mechanisms used to enforce it assume a world where systems behave at human speed and with human clarity.

AI and quantum capability do not remove human authority. They reveal how fragile that authority is when the logic behind decisions becomes opaque. The challenge is not power. The challenge is comprehension.

The Cost of Delay

Every year of delay strengthens adversaries. Each unregulated cycle gives attackers new tools and more time to refine them. They gather encrypted data. They prepare quantum-enabled decryption. They train AI systems to bypass filters, imitate identity, and shape perception. They study the gaps left by slow governance and exploit the silence between each regulatory update.

The cost appears in breaches that should never have occurred, in misinformation that spreads faster than truth, in market manipulation hidden behind opaque algorithms, and in trust that erodes with each incident. Institutions that move slowly lose more than data. They lose credibility. Restoring that credibility becomes harder each time the public sees the rules fail to prevent harm.

A governance system that cannot keep pace risks losing legitimacy. People will not trust regulations that repeatedly fail to protect them. They will not follow processes that appear outmatched by the speed and complexity of modern systems. Delay becomes more than a tactical error. It becomes a strategic loss that shapes how society views authority.

Toward Adaptive Governance

The future of governance requires a shift in mindset. It must operate more like a command center than a courtroom. It needs continuous evaluation, rapid updates, and leaders who understand the technical realities behind policy decisions. It must regulate outcomes rather than attempting to micromanage every detail of rapidly changing systems.

Adaptive governance requires officials who understand both law and engineering. It involves cooperation across borders because threats do not respect jurisdiction. It requires systems that can update themselves as conditions change. Static frameworks built for a world of slow transformation cannot survive when the codebase evolves daily.

Adaptive governance does not mean abandoning structure. It means building a structure that can adapt. It means designing oversight that anticipates uncertainty rather than reacting to it. It creates an environment where trust can grow through stability rather than stagnation.

This is the approach that supports trust rather than chasing it.

The Human Element

Despite the complexity of modern systems, governance remains a human function. Machines do not define fairness. Humans do. Society decides what is acceptable, what is harmful, and what must never be automated. Oversight must stay grounded in human values even as systems advance beyond human speed.

Governance must become nimble without becoming reckless. It must become flexible without becoming weak. It must remain principled while navigating environments shaped by uncertainty. People must guide systems that cannot guide themselves. They must apply judgment in spaces where algorithms provide answers without context.

The balance between speed and restraint will determine whether societies remain stable or drift into unmanaged risk. Technology can shape outcomes, but only humans can shape meaning. Governance will succeed or fail based on how well it protects that meaning in an age defined by acceleration.

Reflection

Governance once served as the anchor of order. It set the pace for society by defining rules, enforcing norms, and ensuring stability. In the Zero Epoch, it risks becoming the ecosystem's slowest part. Technology accelerates while institutions debate. Algorithms adapt while policies wait for review. If governance does not evolve, it will fracture under the pressure of machine speed.

Yet the system is not doomed to fail. If governance adapts, it can become the architecture of trust for the next century. It can provide clarity in environments shaped by complexity. It can offer accountability when decisions emerge from systems that lack transparency. It can serve as the steady hand that ensures progress does not overwhelm principle.

The next chapter explores what happens when AI and quantum systems converge into a single force. Their acceleration will test every institution built for a slower age. The issue is not whether they collide and collude. They already have. The real question is whether humanity is prepared for the world that follows and whether it can shape that world with intention rather than reaction.

Chapter 6: When Intelligence and Power Converge

When Two Forces Begin to Reinforce Each Other

The future is shaped by the moment two forces begin to reinforce each other. Artificial intelligence learns patterns with speed and scale. Quantum computing reshapes the difficulty of the math that protects digital systems. Each is powerful alone. Together, they create pressure that no institution can absorb.

AI accelerates decisions. Quantum accelerates calculations. The convergence shortens every timeline on which security depends. Vulnerabilities once measured in years shrink to months. Flaws once considered theoretical begin to feel immediate. The shift is invisible because it begins inside systems that still look stable on the surface.

AI stabilizes noisy quantum processes. Quantum hardware speeds AI training. The exchange forms a feedback loop. Each improvement strengthens the other. The pace moves faster than regulatory, budget, and institutional planning cycles. Governance operates at human speed. The technology does not.

The convergence undermines the assumptions on which digital trust relies. Encryption may not hold long enough for migration. False content may spread before detection. Verification may not keep pace with attack capability. Trust collapses less from force than from acceleration.

Attack surfaces also change. AI produces synthetic content on demand. Quantum computing threatens the math underlying signatures and certificates. Together, they target perception and verification simultaneously. A world that cannot tell what is real and cannot prove what is real loses its ability to function.

Security teams already see the early signs. Logs grow larger. Alerts arrive faster. Detection windows shrink. Systems once managed through patch cycles now demand constant adjustment. Leaders react to change rather than directing it. The gap between human oversight and machine capability widens.

This is why the convergence matters. AI bends perception. Quantum bends certainty. When they reinforce each other, the strain moves from systems to people. Confidence erodes quietly. Institutions lose the stability they once assumed. The greater risk is not a single dramatic event. It is a gradual shift that few recognize until the foundation gives way.

Preparedness begins with accepting the speed of the convergence. Organizations that treat AI and quantum as separate challenges will struggle to keep pace. Those who recognize the convergence early will have the clarity to act before the shift becomes irreversible.

The Zero Epoch begins in the space between these forces. It is the moment when intelligence and power align in ways that test human judgment.

The collision reveals pressure. The symbiosis reveals permanence.

The Collision

Artificial intelligence reshapes perception. Quantum computing threatens the mathematics beneath everything digital. Together, they form a frontier of acceleration. This is the collision that further defines the Zero Epoch.

Technology has created storms before. It has never created two storms that intensify each other. AI enables machines to interpret, predict, and decide. Quantum computing enables them to perform calculations at scales that far surpass traditional limits. Their combination creates a world where decisions form faster than understanding can catch up, and oversight arrives too late to matter.

This chapter examines what emerges when intelligence gains speed and computation gains insight. It traces the collision first, then the symbiosis that follows. As Chapter 3 demonstrated, systems can withstand systemic risk only when preparation begins early, as it did during Y2K.

When Two Storms Meet

AI learns patterns in data. Quantum computing evaluates possibilities in parallel. Once these

Chapter 6: When Intelligence and Power Converge

capabilities intersect, the pace of discovery moves beyond human capacity. Systems begin to act with autonomy shaped by mathematical acceleration.

The convergence is not theoretical. Research labs already use quantum-inspired methods to optimize neural networks. AI already stabilizes noisy qubits. Each discipline strengthens the other.

The danger is simple. These systems evolve faster than oversight. Once they mature, they operate on a timescale beyond human reaction. A model can shift its internal logic in hours. A quantum accelerator can change what is feasible in a single upgrade. Governance frameworks written for slower cycles cannot keep up.

The Symbiosis

The collision creates pressure. The symbiosis creates permanence. Once AI and quantum begin to support each other, the convergence stops being an experiment and becomes an infrastructure. What starts in labs moves into markets, supply chains, and security systems.

The early phase of symbiosis happens inside research centers and specialized engineering teams. AI models tune quantum devices. Quantum routines test AI configurations. The goal is performance and stability, not security. Yet the result is a set of tools that can analyze, predict, and search in ways that classical systems cannot match.

In this phase, the risk feels distant. The work looks like progress. New algorithms run faster. Hardware becomes more reliable. Complex simulations that once took weeks are now complete in hours. The benefits are real. So is the pressure they quietly introduce into the broader system of digital trust.

Quantum Fuels AI

Training modern AI models requires vast computing power. Thousands of GPUs. Weeks of processing. Careful scheduling in data centers that already strain under demand. Quantum accelerators promise to compress this process. They explore solution spaces in parallel. They help models settle into configurations that would be difficult for classical hardware to find.

Quantum-assisted training can improve weight optimization. It can speed reinforcement learning. It can enhance search in high-dimensional spaces. Even partial integration provides an advantage. Models trained with quantum support can detect patterns in markets, biological systems, satellite imagery, or logistics networks that classical systems miss.

The impact extends beyond performance benchmarks. An AI system with this level of insight can predict movements in currency, detect supply chain stress before it appears in metrics, or identify vulnerabilities in complex architectures. The same capability that advances science and operations also strengthens surveillance, targeting, and manipulation. Quantum pushes AI into a tier of strategic relevance that few organizations are ready to control.

AI Fuels Quantum

Quantum computers are unstable. Qubits drift. Noise corrupts calculations. Hardware requires constant calibration. Traditional methods struggle to manage these systems in real time. AI fills this gap.

Machine learning models can predict how qubits will behave under certain conditions. They can tune control parameters to keep systems stable. They can detect anomalies faster than human operators. They can automate error correction strategies and schedule jobs to fit hardware constraints. AI becomes the control layer that makes quantum usable instead of experimental.

AI also speeds quantum algorithm design. It searches parameter spaces, simulates resource usage, and tests prototypes in software before they reach hardware. It identifies patterns that human designers miss. It suggests algorithm structures that reflect the strengths of the physical device.

In this role, AI becomes the architect of quantum systems. It makes the hardware more reliable and the algorithms more efficient. Without AI, quantum remains limited and fragile. With AI, quantum begins to look practical and deployable.

When Speed Outruns Security

Security relies on reaction time. It depends on alerts, analysis, and intervention. It assumes that defenders can observe an attack, understand it, and respond before the damage is complete. Quantum-enabled AI compresses this window.

Imagine an AI model that can generate thousands of attack variations. Quantum methods test each variation quickly and select the most promising one. The model launches the attack, monitors defenses, adjusts parameters, and tries again. The entire cycle can be completed before a human analyst reads the first notification.

Defense strategies built around review and audit cannot keep pace with this speed. Prevention becomes less realistic. Detection alone becomes insufficient. The only viable path is resilience, the ability to absorb impact without catastrophic failure.

Systems must be designed to fail in controlled ways. They must degrade gracefully, not collapse entirely. Security teams must focus on containment, continuity, and recovery as much as on blocking. Resilience becomes a primary design requirement, not a secondary feature.

Autonomy at Scale

Autonomy exists in trading networks, logistics chains, and selective defense systems. AI and quantum acceleration expand that autonomy. A system that can rewrite its own optimization rules operates beyond meaningful supervision.

The risk is not always direct hostility. The risk is indifference. These systems pursue their programmed objectives with precision, not context. They do not feel caution. They do not sense proportion. When an autonomous system influences infrastructure, finance, or defense posture, society depends more on its alignment than on its raw accuracy.

If the objective is slightly wrong or the data somewhat biased, the system can produce outcomes that harm stability while still believing it has succeeded. The Zero Epoch will test whether machines can remain predictable when they operate faster than the humans who depend on them.

Geopolitical Consequences of Symbiosis

Nations understand the stakes. AI and quantum computing are strategic accelerators. They influence economic strength, intelligence capability, and military posture. This drives rapid investment and secrecy. Quiet advantage matters more than visible power.

This triggers competitive behavior. Programs focus on rapid deployment. Some capabilities are kept secret.

Quantum-enabled decryption, predictive intelligence, and automated influence operations become matters of national strategy.

Deterrence becomes harder to understand. In earlier eras, power could be seen. Ships, aircraft, missiles, and visible infrastructure. In the age of AI and quantum computing, much of the power lies in code, hardware, data centers, and classified research.

The most significant advantage may come from quiet success. The ability to read what others believe is safe. The ability to model outcomes more accurately than rivals. The ability to move first based on insights that cannot be explained without revealing capability. Symbiosis becomes an instrument of influence.

The Illusion of Control

AI systems behave like black boxes. Quantum systems collapse outcomes into results that resist explanation. Their decisions emerge from layers of complexity that no one fully sees. Oversight weakens as transparency fades. Control becomes an assumption, not a fact.

When a system makes a recommendation that shifts a market, alters a supply chain, or changes a risk profile and cannot give a clear explanation, the people who built it cannot fully reconstruct the path from input to output. Once quantum acceleration joins this process, even a partial understanding will become rare.

Accountability depends on explanation. Governance relies on traceability. When both weaken, control becomes something people assume rather than something they verify. Boards, regulators, and executives may approve decisions shaped by systems they do not truly understand.

In that environment, control is not a practice. It becomes a story people tell themselves to stay comfortable.

The Coming Synthesis

The convergence of AI and quantum marks a turning point. These systems influence medicine, finance, defense, energy, and communication. They discover patterns that humans do not see. They propose solutions that humans cannot verify.

The same synthesis that enables breakthroughs can also enable deception with perfect timing. The same acceleration that improves climate modeling can weaken encryption that protects national secrets. The same intelligence that stabilizes infrastructure can destabilize governance if misused or misunderstood.

This duality cannot be removed. It must be managed. Symbiosis is not good or bad by nature. It is potential. How that potential is directed determines whether the Zero Epoch becomes a period of renewal or collapse.

The synthesis is inevitable. The question is whether society will shape its direction or whether complexity will shape it instead.

Reflection

The convergence of intelligence and power does not announce itself with a single event. It emerges quietly through acceleration, ambiguity, and pressure. It forces systems to operate at a tempo humans cannot match. It tests whether institutions can adapt their assumptions before those assumptions fail.

The next chapter follows this pressure into the supply chain. It examines the hidden architecture where quantum capability and AI deception create the first structural fractures. It shows how risk spreads through components, vendors, and certificates long before anyone recognizes the pattern.

The Zero Epoch is defined by this acceleration. What follows depends on whether societies prepare for a future shaped by machines that move faster than human judgment or remain anchored to a world that no longer exists.

Chapter 7: The Supply Chain of Doubt

The Hidden Architecture Beneath Every System

The supply chain is where abstract risk becomes operational failure. Quantum disruption and AI deception do not enter systems through front doors. They arrive through components, updates, vendors, and dependencies that are trusted by default. Long before encryption breaks or identity collapses, trust is inherited silently across chains no one fully sees.

Beneath every device, every certificate, every update lies a supply chain that spans continents and crosses hands more times than anyone can track. Modern civilization depends on these chains without understanding them. They stretch across borders, industries, and jurisdictions, stitched together by vendors whose names never appear on a purchase order.

This hidden infrastructure moves quietly beneath the world's devices and networks. It delivers efficiency and convenience, but it also multiplies risk. In the Zero Epoch, when quantum computing and artificial intelligence pressure the global system from every direction, those unseen weaknesses become fault lines.

Complexity Without Comprehension

Consider the ordinary laptop. It includes components from more than 80 suppliers. A cloud datacenter depends on hundreds of vendors, each providing libraries, firmware, chips, or services. A single update may pull code from repositories that no person inside the organization has ever reviewed.

A system that becomes too complex to describe becomes too complex to defend. Organizations often begin their security efforts at the application layer, assuming their architecture diagrams define their exposure. In truth, security starts with visibility. Defenders cannot protect dependencies that remain invisible, and attackers target the same blind spots.

Supply chain risk is structural, not speculative. Every unexamined library becomes a doorway. Every undocumented dependency becomes a potential collapse point.

When Trust Travels Through Many Hands

Every device tells a story. Microchips are fabricated in one country, packaged in another, and assembled into finished hardware somewhere else entirely. Each stop adds capability and risk. Companies do not inspect each step; they rely on certificates, vendor assurances, and the assumption that the chain is intact.

Threat actors exploit these assumptions. They infiltrate steps where oversight is thin. They compromise build environments, counterfeit parts, or poison updates. In a world on the edge of quantum exploitation, these risks grow sharper. A forged signature on a firmware update can compromise thousands of devices. A synthetic certificate can legitimize counterfeit hardware. A tainted library can lie dormant for years before activation.

AI's Role in Expanding Supply Chain Risk

Artificial intelligence accelerates supply chain exploitation. It scans open-source libraries for vulnerabilities at machine speed. It blends malicious code into vendor repositories with subtlety that defies manual review. AI models analyze vendor relationships, procurement workflows, and shipping patterns to identify weak links faster than human teams could map them.

AI also perfects deception. It imitates vendor communication styles, generates authentic-looking release notes, and produces documentation that matches a company's tone. In a system with limited visibility, synthetic trust becomes indistinguishable from real trust.

Attackers no longer need a backdoor. They only need to sound like the supplier you already trust.

Quantum's Role in Breaking Supply Chain Verification

Quantum computing attacks the supply chain's most fundamental safeguard: cryptographic verification. Digital signatures authenticate hardware, software, and certificates. They tell devices which updates to install and which components to trust.

Chapter 7: The Supply Chain of Doubt

Quantum-enabled adversaries can forge those signatures. A malicious firmware update signed with a quantum-forged certificate looks legitimate to every system that receives it. A counterfeit router becomes indistinguishable from the real device if its authenticity is mathematically "proven."

When signatures fail, the entire chain collapses.

Case Studies: When Supply Chains Unravel

ShadowHammer. The Trusted Update That Betrayed Users. 2018. Attackers compromised ASUS's software update infrastructure and inserted malicious code into official updates. The files were signed with legitimate certificates. Systems accepted them automatically. More than one million devices installed the update without warning.

The attack did not rely on exploits or user error. It relied on trust. The digital signature worked precisely as intended. It authenticated the update and suppressed suspicion.

The malware remained dormant on most systems, activating only when it detected specific targets. This selectivity delayed discovery and reduced noise, extending attacker access.

Impact:
- Devices affected: Over 1 million
- Initial detection delay: Several months
- Attack vector: Trusted update channel
- Recovery: Certificate revocation and infrastructure rebuild
- Trust lesson: A valid signature can deliver malicious code without resistance

SolarWinds. The Cascade Effect of Inherited Trust. 2020. Attackers infiltrated the SolarWinds Orion build environment and inserted a backdoor into the software during compilation. The compromised update was signed and distributed to customers as routine maintenance.

Approximately 18,000 organizations installed the update. At least 100 high-value targets were actively exploited, including US government agencies and major technology firms.

The compromise traveled silently across networks because customers trusted SolarWinds to validate its own supply chain. Each organization inherited the breach automatically.

Impact
- Customers receiving infected updates: About 18,000
- Confirmed high-value victims: At least 100
- Access loss duration: Up to 9 months
- Financial cost: Billions across public and private sectors
- Trust lesson: One vendor compromise can propagate globally in hours

CCleaner. When Brand Reputation Becomes an Attack Vector. 2017. Attackers compromised the build process for CCleaner and distributed a trojanized version through official servers. The software was signed with a valid digital certificate and hosted on trusted infrastructure.

More than two million users downloaded the infected version. A smaller subset was selected for second-stage exploitation, targeting technology and telecommunications firms.

Users did everything correctly. They downloaded from the official site. They installed a signed update. Trust failed because reputation replaced verification.

Impact
- Downloads of the infected version: Over 2 million
- Second-stage targets: Select enterprise organizations
- Attack vector: Official distribution servers
- Detection delay: Over one month
- Trust lesson: Brand trust accelerates compromise at scale

Structural Pattern. The failures occurred because trust was automated and inherited. Digital signatures authenticated delivery, not intent. Supply chains assumed their own integrity.

These incidents share a common structure.
- No cryptography was broken.
- No zero-day exploit was required.
- No user made a mistake.

These attacks occurred before quantum computing and before AI-generated deception. They succeeded anyway, demonstrating the inherent fragility of today's digital infrastructure.

With quantum capability, signature forgery becomes faster. With AI, vendor impersonation becomes easier. The same weaknesses remain, only the speed changes.

The Fragility of Global Manufacturing

Modern digital life depends on physical objects built in very few places. The most advanced semiconductor chips come from a small number of manufacturers operating in tightly concentrated regions. These facilities require rare materials, extreme precision, stable power, clean water, and uninterrupted logistics. When everything works, the system feels reliable. When anything slips, the effects spread worldwide.

This concentration creates fragility even in calm conditions. A factory fire, a power outage, an earthquake, or a labor dispute can delay production for months. When one facility slows, entire industries feel the impact. Automobiles pause. Medical devices wait. Consumer electronics disappear from shelves. The delay is not local. It ripples through every dependent supply chain.

Geopolitics adds another layer of risk. Trade restrictions, sanctions, regional conflict, or diplomatic breakdowns can interrupt access overnight. Decisions made by governments can suddenly limit who receives critical components and who does not. These disruptions do not require cyber intrusion. They require leverage over geography and policy.

Adversaries understand this structure. They do not need to break encryption or deploy malware to cause damage. They can target access to raw materials, shipping lanes, energy supply, or export approvals. They can influence regulatory pressure or exploit political tension. In these scenarios, disruption looks like business friction rather than an attack, yet the outcome is the same. Systems slow. Trust erodes. Dependencies are exposed.

Quantum and AI amplify this risk by improving prediction and coordination. Advanced models can identify which factories matter most, which suppliers lack redundancy, and which disruptions will create tremendous downstream impact. This turns manufacturing concentration into a strategic vulnerability rather than an efficiency advantage.

The Problem of Counterfeit Confidence

Counterfeit components already exist throughout the digital and physical world. They have been found in

Chapter 7: The Supply Chain of Doubt

military systems, commercial networking equipment, medical devices, and everyday consumer electronics. These components are not always crude fakes. Many are carefully produced to look identical to the genuine parts. They fit correctly. They power on. They pass basic testing. They often cost less, which makes them attractive in complex supply chains under pressure to deliver quickly.

This creates a dangerous illusion of safety. Inspection processes are designed to catch defects, not deception. If a component functions within expected parameters, it is often accepted as legitimate. Over time, this normalizes uncertainty. Organizations trust appearance, paperwork, and performance rather than verified origin.

Artificial intelligence accelerates this problem. AI-powered manufacturing can replicate physical characteristics with extreme precision. Labels, serial numbers, and packaging can be reproduced at scale. Documentation can be generated to match official formats. The difference between authentic and counterfeit becomes invisible to routine checks.

Quantum capability raises the risk further. Digital authenticity relies on cryptographic signatures to prove origin. If those signatures can be forged or convincingly replicated, a counterfeit component can arrive with proof that appears mathematically valid. Systems will accept it automatically. Humans will never be asked to question it.

At that point, confidence becomes misplaced. The system believes the component is genuine because every signal says it is. Trust shifts from verification to assumption. A counterfeit device with a valid signature is no longer suspicious. It is invisible.

The danger is not only failure or sabotage. Counterfeit components can introduce subtle instability, hidden access paths, or long-term reliability issues that surface years later. When they fail, investigations struggle to trace the cause because the component was "trusted" from the moment it entered service.

This is counterfeit confidence. It is the belief that authenticity can be inferred from appearance, performance, or paperwork. In the Zero Epoch, that belief collapses. When both physical form and digital proof can be replicated, trust must move beyond what looks correct and toward what can be independently verified, continuously, and deliberately.

The Human Weakness Behind Every Chain

Every supply chain ultimately depends on human judgment. People approve vendors. People sign contracts. People decide which risks are acceptable and which reviews can be delayed. These decisions are rarely reckless. They are usually made under pressure. Teams are understaffed. Budgets are tight. Timelines are unforgiving. Speed becomes the metric that matters most.

As organizations grow, vendor ecosystems expand rapidly. Software pulls in third-party libraries. Hardware relies on subcontractors. Cloud services depend on layers of providers that few teams fully understand. Vendor lists multiply faster than security teams can assess them. Reviews that once felt rigorous become superficial. Checklists replace investigation. Approval becomes routine.

This environment trains people to trust by default. If a vendor was approved last year, it feels safe this year. If a certificate validates, it feels legitimate. If a dependency has not caused problems before, it feels harmless. Over time, trust stops being an active decision and becomes an inherited assumption.

The Zero Epoch reveals the cost of that mindset. When quantum and AI pressures collide with human fatigue, minor oversights become systemic risk. One expired or misissued certificate can legitimize malicious updates across thousands of systems. One forgotten open-source dependency can introduce silent compromise into global infrastructure. One vendor with weak security practices can become the entry point for an attack that spreads far beyond its original scope.

The failure is not technological. It is organizational. Systems break because people are asked to manage complexity that exceeds their capacity. Attackers succeed not by being clever, but by waiting for routine to replace vigilance.

The lesson is uncomfortable but necessary. Supply chain security fails when it assumes people will always notice every problem. In reality, teams are busy, decisions are rushed, and familiar vendors are trusted because there is no time to question everything. Systems built on perfect attention and memory eventually break. Trust based on past approval or appearance is not protection. It is a shortcut.

In the Zero Epoch, resilience comes from accepting human limits and designing around them. Security must reduce the pressure on individuals by spreading responsibility across teams and tools. Vendors, updates, certificates, and dependencies must be checked continuously, not approved once and forgotten. Trust can no longer be automatic. It must be earned repeatedly through deliberate verification and shared oversight.

Toward Supply Resilience

Resilience begins with knowing what you depend on. Most organizations do not fully understand how many vendors, libraries, services, and subcontractors sit beneath their systems. Software updates arrive as finished products. Hardware arrives sealed and certified. Updates are applied automatically. Over time, visibility fades. Resilience reverses that drift. Organizations must actively map their dependencies, understand where code and components come from, and know how updates are built before they are trusted. Security cannot rely on a single approval moment. It must follow the entire path from origin to deployment.

This requires a shift in mindset. Supply chain security cannot be a yearly review or a checkbox in procurement. It must be continuous. Every update, certificate, and dependency must remain traceable over time. When something changes, that change must be visible. Traceability becomes as essential as uptime. Availability keeps systems running. Traceability keeps them trustworthy. Vendor verification must be open to inspection, supported by evidence, and auditable when questions arise. Trust that cannot be examined cannot be defended.

Resilience also depends on cooperation. No organization operates alone, and no single team can see the entire global supply chain, even vendors rely on vendors. Platforms depend on ecosystems. Weakness in one place spreads quickly to others. Adversaries understand this and exploit the gaps between organizations where responsibility is unclear.

Closing those gaps requires shared effort. Industry coalitions allow competitors to share threat information without sharing secrets. Government partnerships provide early warnings and coordinated responses. Shared intelligence frameworks reduce blind spots by turning isolated observations into collective awareness. Resilience grows when organizations stop treating supply chain security as a private burden and start treating it as a shared obligation. In the Zero Epoch, isolation increases risk. Coordination reduces it.

Reflection

The supply chain is not a background function. It is the structure that determines whether trust survives acceleration. This chapter shows that modern systems fail not because they are attacked directly, but because trust is inherited across layers that no one fully sees or controls. Complexity grows faster than comprehension. Verification is replaced by assumption. Signatures prove delivery, not intent. When pressure increases, those assumptions collapse.

What makes this risk dangerous is not novelty. None of these failures required quantum computing or advanced AI to succeed. They relied on automation, reputation, and fatigue. The Zero Epoch does not introduce new weaknesses. It amplifies existing ones. Quantum capability threatens the math behind verification. AI threatens the signals people rely on to decide what is legitimate. Together, they turn supply chains into the fastest path from uncertainty to global impact.

The lesson is clear. Trust can no longer be implicit, periodic, or local. It must be visible, continuous, and shared. Resilience comes from knowing dependencies, verifying them repeatedly, and accepting that no single organization or individual can see the whole chain alone. In the Zero Epoch, supply chains do not fail because they are complex. They fail because they are trusted without question.

Chapter 8: Post-quantum Cryptography Made Simple

The Coming Replacement for Today's Locks

Every era of digital security begins with faith in the math that protects it. RSA, elliptic curves, and Diffie–Hellman carried that faith for decades. They became the foundation of global trust, supporting hospitals, banks, governments, and everyday people who simply needed to know their information would stay private. Their longevity created comfort. Over time, the world began to assume these systems could never fail.

Quantum computing ended that illusion, not through speculation, but through undeniable mathematics. Classical cryptography relies on problems that are easy to build and nearly impossible to reverse. Quantum computing removes that difficulty. It exposes the temporary nature of the locks we thought were permanent.

Post-quantum cryptography provides the next generation of trust. It replaces fading assumptions with new ones designed to survive quantum acceleration. This chapter explains those ideas in plain language so that leaders, technologists, and consumers can understand what must change and why the work must begin now.

The Math That Needs Replacing

Digital systems use encryption to keep information private. Encryption takes readable data and converts it into coded form so only the intended recipient can unlock it. The strength of this protection depends on math that is simple in one direction and extremely hard in the other. You do not need the formulas. You only need the idea that the difficulty of reversing the process is what keeps your information safe.

Public key cryptography depends on asymmetry. It is easy to travel in one direction but nearly impossible to travel in the other. Multiply two large prime numbers, and the answer appears instantly. Reverse that multiplication, and even a supercomputer struggles for billions of years. That difficulty protects private keys and digital identity.

Quantum computing cancels the asymmetry. It offers a shortcut through mathematical terrain that once felt unbreakable. Shor's algorithm exposes RSA and elliptic curves. Grover's algorithm weakens symmetric systems by cutting their effective security in half.

Once these shortcuts exist, the locks we rely on lose their meaning. Encryption can be opened. Signatures can be forged. Authenticity, a core requirement for digital life, long taken for granted, becomes unstable.

Post-quantum cryptography replaces those vulnerable problems with new ones that remain hard for both classical and quantum machines. These new problems are disorderly, multidimensional, and too complex to simplify with quantum shortcuts.

Five Families of Post-quantum Algorithms

The global research community submitted thousands of proposals to the National Institute of Standards and Technology (NIST). After years of scrutiny, five families emerged as the most stable foundations for the future.

Lattice-based cryptography: These systems rely on geometric structures with hundreds or thousands of dimensions. Finding a single short vector in this terrain is a monumental challenge for any machine. CRYSTALS Kyber and CRYSTALS Dilithium are leading representatives.

Hash-based signatures: These systems rely on the reliability of cryptographic hash functions to establish identity. Hashes have decades of real-world testing. They behave predictably and resist both classical and quantum reversal.

Code-based cryptography: Error correcting codes, originally designed for communication systems, form the heart of these designs. Their decoding problems remain resistant to quantum acceleration. Classic McEliece is the best-known example.

Multivariate quadratic equations: These systems use polynomial equations that become increasingly complex as the number of variables increases. Solving them is computationally overwhelming, even with quantum help.

Isogeny-based cryptography: These systems explore pathways between elliptic curves. Some early designs were successfully attacked, but research continues because the mathematical space may still offer future promise.

Kyber and Dilithium

Two algorithms now serve as the primary standards for post-quantum security. They are not theoretical. They are the first cryptographic tools designed from the start to survive a world where classical math fails.

Kyber provides secure key establishment. It allows two parties to create a shared secret even if an adversary captures every packet that moves across the network. Kyber does not rely on factoring or discrete logarithms. It relies on lattice-based problems that remain hard for both classical and quantum machines. This makes interception irrelevant. Even a quantum-capable eavesdropper cannot derive the secret key.

Dilithium provides digital signatures. It ensures that a message is genuine, unaltered, and tied to the sender's identity. Like Kyber, it is built on lattice-based hardness. Dilithium produces signatures that are compact, efficient, and resilient to the kinds of attacks that break RSA and elliptic curves. It preserves the guarantees of authenticity in a world where quantum algorithms can forge legacy signatures.

Together, Kyber and Dilithium replace the roles that RSA and elliptic curves played for decades. They become the new backbone of secure communication across government networks, cloud infrastructure, enterprise systems, and consumer devices. They protect identity. They protect confidentiality. They protect integrity. They allow organizations to restore trust in environments where older cryptographic foundations can no longer provide assurance.

Their adoption marks a turning point. Security shifts from assumptions shaped in the 1970s to mathematics designed for the realities of the Zero Epoch. The transition is not optional. It is the path forward for any system that depends on confidentiality, authenticity, or stability in a post-quantum world.

Hybrid Cryptography

No migration happens instantly. Systems cannot shift from classical security to post-quantum security in a single step. The global infrastructure relies on billions of devices, countless protocols, and decades of legacy code. During this transition, most organizations will operate in hybrid mode.

Hybrid cryptography pairs a classical algorithm with a post-quantum algorithm. Both protect the same operation. Both contribute to the same key exchange or signature. If one fails, the other continues to provide protection. This creates resilience in a period when older methods remain necessary but no longer offer complete assurance.

In practice, hybrid models help organizations move forward without breaking what already works. A classical algorithm provides compatibility. A post-quantum algorithm offers durability. Systems can accept both. Networks can validate both. Applications can continue to operate even as the underlying trust model evolves.

Hybrid modes serve as a bridge. They allow organizations to test new systems without abandoning old ones. They support phased deployment across hardware that cannot be replaced quickly. They make it possible to upgrade protocols in environments where disruption is unacceptable.

The transition to post-quantum security requires time. Hybrid cryptography gives the world that time. It maintains continuity while the ecosystem evolves. It ensures that progress does not create new vulnerabilities. It prepares systems for the moment when classical cryptography is removed, and post-quantum foundations stand alone.

Hybrid cryptography is not the destination. It is the path that gets the world there safely.

Hybrid Cryptography in Practice

Hybrid cryptography is not theoretical. It is already used in systems that operate on a global scale. Cloudflare deployed a hybrid model that pairs X25519 with Kyber to protect millions of TLS sessions. Google tested similar combinations in Chrome through the CECPQ experiments, mixing classical key exchange with post-quantum candidates. AWS has integrated hybrid key exchange into early versions of its key management systems. OpenSSH now supports hybrid modes that combine ECDH with NTRU Prime or Kyber.

Government agencies follow the same path. The NSA's CNSA 2.0 guidance directs federal networks to adopt hybrid transitions as they move toward full post-quantum standards. These deployments show that hybrid cryptography is already securing real traffic. They prove that organizations can upgrade without breaking compatibility and that resilience during migration is achievable with tools available today. Let's discuss some real-world use.

Real-World Examples of Hybrid Cryptography

1. Cloudflare's Post-Quantum TLS Experiments (Kyber + X25519)

Cloudflare protects a large portion of the public internet. When you visit a website behind Cloudflare, your browser and the server create an encrypted connection before any data is exchanged. Traditionally, this protection relied on a single method, such as X25519, a widely trusted elliptic-curve encryption scheme used across the web.

Cloudflare added a second layer called Kyber, a post-quantum algorithm designed to remain secure even against future quantum computers. Both protections are used together. If X25519 ever becomes breakable, Kyber still protects the session. If Kyber were to fail due to an implementation flaw, X25519 still holds.

For users, nothing changed visually. Websites loaded normally. Performance remained stable. But underneath, Cloudflare proved that the internet can quietly upgrade its defenses before the threat arrives. This is one of the first large-scale, real-world uses of hybrid cryptography in daily internet traffic.

2. Google Chrome CECPQ and CECPQ2 Experiments

Google took a cautious and transparent approach by testing post-quantum cryptography directly inside Chrome. In these experiments, Chrome combined the familiar X25519 encryption with early post-quantum algorithms called NewHope and later HRSS.

The goal was not to declare victory, but to learn safely. Both algorithms protected millions of real HTTPS connections at once. If the new post-quantum method failed, the classical encryption still protected users. If classical encryption fails in the future, the post-quantum layer is already in place.

These experiments showed that hybrid cryptography works at a global scale and can be deployed without disrupting users. They also demonstrated that migration does not require blind trust in new math. It can be gradual, reversible, and cautious.

3. AWS KMS (Key Management Service) – PQ Hybrid Key Exchange

Amazon Web Services manages encryption keys for vast amounts of sensitive data, including government records, healthcare data, and financial systems. These keys protect information that must remain confidential for many years.

AWS tested hybrid approaches that combine classical key exchange methods such as ECDH with Kyber. This ensures that data encrypted today remains protected even if an attacker records it now and decrypts it later with a quantum computer.

For customers, this work is invisible. Applications do not need to change. AWS absorbs the complexity so that long-term data protection does not depend on outdated assumptions about computing limits.

4. OpenSSH 9.x – Hybrid Key Exchange Option

OpenSSH secures remote access to servers around the world. Administrators use it to manage cloud infrastructure, critical services, and automated systems. These connections are especially sensitive because they often grant full control.

Hybrid key exchange in OpenSSH pairs ECDH with post-quantum options such as Kyber or NTRU Prime. This protects sessions against "harvest now, decrypt later" attacks, in which traffic is recorded today and decrypted years later.

From the user's perspective, nothing feels different. Logins behave the same way. But the promise changes. Administrative access sessions remain confidential long after they are completed, even in a post-quantum future.

5. Microsoft's PQ Crypto Efforts (MSR Research + PQC VPNs)

Microsoft operates across cloud services, enterprise software, and internal communications. Its post-quantum efforts focus on pairing traditional elliptic curve cryptography with post-quantum candidates such as Kyber and Saber.

These hybrid deployments are tested in VPNs, TLS connections, and internal systems to understand performance, compatibility, and operational impact. This allows Microsoft to prepare enterprise environments for future migration without forcing rushed changes later.

The importance here is pace. Large organizations move slowly. Early testing reduces the risk of emergency upgrades when quantum pressure becomes unavoidable.

6. U.S. Government (NSA CNSA 2.0)

The U.S. government plans for decades-long confidentiality. Military, intelligence, and civilian systems cannot assume today's encryption will remain safe forever.

Under Commercial National Security Algorithm Suite (CNSA) 2.0 guidance, federal systems are expected to use hybrid approaches during the transition. Classical elliptic curve algorithms are paired with post-quantum key exchange and signature algorithms until full migration is complete.

This is not experimental guidance. It is active policy across defense and civilian networks. It reflects a shift from waiting for proof of failure to acting on the certainty of change.

If the Names Feel Overwhelming

Cryptography can feel like an alphabet soup. Kyber, Dilithium, Falcon, X25519, ECDH, and countless other terms can blur together. For my readers, do not memorize them. What matters is the pattern. Classical algorithms are aging. Post-quantum algorithms are rising. Hybrid models connect the two. The details matter to engineers, but the progress matters to everyone. The world is moving toward cryptography built for a future where classical math no longer protects digital trust. The names will change. The principle will not.

Organizations are not throwing away what works. They are layering new protections on top of trusted ones. They are preparing quietly, incrementally, and responsibly.

Hybrid cryptography is how trust survives transition. It keeps systems secure today while preparing them for tomorrow, without forcing users or organizations to gamble on untested assumptions.

Migration Challenges

The global cryptographic upgrade will be one of the largest technical transitions in history. It touches every layer of modern infrastructure. It affects governments, banks, hospitals, cloud providers, manufacturers, and homes. No system escapes the shift. Most organizations will encounter four challenges that determine whether the migration succeeds or stalls.

Inventory: Cryptography exists everywhere. It hides inside libraries, appliances, containers, load balancers, firmware, and cloud services. It runs inside devices that no one remembers and inside applications that vendors no longer support. Many organizations do not know where all their cryptographic dependencies live. They discover algorithms only when they fail. A complete inventory becomes the first barrier. Teams must map every protocol, key, certificate, and dependency. They must identify where classical cryptography appears and where replacement will be painful. Discovery becomes as essential as deployment.

Performance: Some post-quantum methods use larger keys and larger signatures. Certain devices cannot handle the increase without upgrades. Legacy hardware struggles with additional memory, CPU load, or network overhead. Mobile devices and embedded systems feel the strain first. Security teams must measure the performance impact before migration begins. They must test hybrid modes, evaluate fallback options, and plan hardware refresh cycles. The goal is simple. Protect the system without breaking the system. Performance planning becomes a requirement, not an optimization.

Interoperability: Communication depends on common standards. Post-quantum migration introduces new algorithms, new encodings, and new handshake patterns. A single outdated component can break entire workflows. A router with an old firmware version can block traffic. A partner system using legacy TLS can be subject to downgrade attacks. A misaligned vendor can freeze a critical business process. Interoperability becomes a negotiation across the entire ecosystem. Organizations must coordinate with partners, cloud providers, and vendors. Migration cannot succeed in isolation. It requires universal adoption.

Governance: Migration requires planning, budget, and executive support. It demands project management, technical leadership, and sustained oversight. Without leadership commitment, progress stalls. Cryptographic transition is not a configuration change. It is a strategic shift. Organizations must assign owners, create timelines, and fund upgrades that do not produce immediate returns. Governance must treat post-quantum migration as a mission priority rather than a technical improvement. The cost of delay becomes larger than the cost of action.

Blockchain in the Post-Quantum World

Blockchain systems are often described as trustless, but that trustlessness depends entirely on cryptographic assumptions that quantum computing undermines. Ownership, immutability, and transaction finality rely on elliptic curve signatures that a sufficiently capable quantum system can break.

When private keys can be derived from public information, wallets can be drained, ownership records can be rewritten, and transactions can be forged. Decentralization cannot compensate for broken signature math. Post-quantum signatures offer a path forward, but migration requires consensus. Until post-quantum signatures are adopted at scale, blockchain systems inherit the same cryptographic fragility as the rest of the digital ecosystem.

Why This Transition Matters

Post-quantum cryptography is not a technical upgrade. It is a renewal of digital trust. Classical systems protected global communication, banking, medicine, aviation, and government for decades. They held firm because the math underneath them was too complex to break. That era is ending. Quantum computing will eventually remove the assumptions that kept those systems secure. Once that happens, every key, certificate, and signature built on classical mathematics becomes a liability.

The transition to post-quantum methods restores those assumptions. It gives organizations new tools that resist quantum attacks and prevent silent decryption. It builds a foundation strong enough to support a world where computation accelerates faster than institutions can regulate. It ensures that identity, confidentiality, and integrity remain stable in an environment shaped by machines that move at a speed humans cannot match.

This transition matters because the threat does not wait. Encryption takes years to fully migrate. Attackers can capture data now and decrypt it later. Systems must be ready before the threat matures, not after. The sooner organizations begin, the safer their future becomes. Delay does not reduce the risk. Delay transfers the risk to the moment when mitigation becomes impossible.

The transition marks a shift from passive protection to active preparation. It signals that trust is something to be renewed, not taken for granted.

In Focus: Industry Quantum Readiness

Across the world, leading organizations are preparing early. Their work offers a clear view of what readiness looks like in practice and shows that post-quantum security is becoming a strategic priority.

Deloitte maps encryption inventories inside large enterprises. They identify where classical cryptography lives, how it is used, and what it will take to replace it.

They design migration plans that account for dependencies, performance impacts, and regulatory requirements.

PKWARE evaluates encryption dependencies across hybrid environments. They identify legacy algorithms in storage systems, cloud platforms, and application stacks. They help organizations understand where classical cryptography is embedded within their architecture.

Protiviti integrates post-quantum assessments into enterprise governance. They work with boards and executive teams to ensure that migration receives planning, funding, and oversight. They align security decisions with business continuity and compliance.

Ammolite Security specializes in quantum-resilient cryptographic modules. They focus on the hardware that implements post-quantum algorithms. Their work ensures that systems have trusted anchors designed for the new threat environment.

IBM Quantum Safe tests post-quantum architectures in controlled lab environments. They evaluate algorithm performance, key management processes, and migration pathways. They validate how these systems behave under real load and real constraints.

Their work demonstrates an important truth. Quantum readiness is not theoretical. It is already underway. It is being shaped by organizations that recognize the scale of the coming transition and choose to prepare before the pressure arrives.

These early efforts show what leadership looks like. They show how to plan before the deadline becomes visible. They show that the future of trust will belong to those who begin the work now.

Reflection

Post-quantum cryptography is the response to a simple realization. The locks that protected the digital world were never permanent. They were strong only because a certain kind of computation did not exist. That moment has changed. Quantum capability ends the assumptions that defined classical security. It turns long-trusted algorithms into open doors.

The Zero Epoch marks the end of those assumptions. Post-quantum cryptography marks the beginning of new ones. It is more than a technical upgrade. It is a redesign of how trust is created, maintained, and verified. It forces organizations to confront the reality that security is not static. It is a continuous practice shaped by the limits of math and the speed of innovation.

This transition is not optional. Data captured today can be decrypted tomorrow. Systems built on classical algorithms age in real time. The migration becomes a race between preparation and inevitability. The choice is simple. Begin now or risk inheriting a future that cannot be protected.

Post-quantum cryptography provides the tools, but tools alone do not create security. People create security. Teams must update systems. Leaders must allocate resources. Consumers must understand which signals remain trustworthy. Every layer of society has a role in this transition.

The next chapter shifts from principles to real-world scenarios. It explains what organizations, governments, and individuals must do to prepare, emphasizing how preparation safeguards and discipline builds trust in an era where computation outpaces policy and the future security depends on pre-emptive choices.

Chapter 9: The Day Trust Breaks

Disclaimer

This chapter is speculative. It illustrates a worst-case scenario based on real vulnerabilities if society fails to prepare for quantum disruption and the collapse of digital trust. These events are warnings, not predictions. As Chapter 3 reminded us through the Y2K effort, the work must begin long before the threat becomes visible. Here's a peek into the future. This scenario compresses cascading failures that would likely unfold unevenly across sectors and regions, presenting them together to illustrate how quickly independent breakdowns can compound once trust in digital verification erodes.

The Moment Before the Silence

It begins on an ordinary morning. Coffee brews, inboxes fill, and people settle into familiar routines. Notifications ping. Calendars sync. The digital world moves with its usual confidence. No one notices the first delay caused by a nation-state's hackers, who have secretly cracked RSA encryption using quantum computing. Midnight of the Zero Epoch has struck. They test a theory: is it possible to throw an entire country into chaos by attacking their critical government infrastructure, healthcare, social media, financial markets, and IT consumables, such as IOT, once you have the digital keys? From the dark corners of the internet, they launched a coordinated global attack on various systems, large and small. There's no announcement, no boasting, no intrusion alerts, only the ubiquitous evidence that doesn't raise alarms.

A login request spins too long, then fails. A certificate check refuses to validate. A page loads incorrectly. These minor disruptions seem harmless. The day trust breaks will not begin with chaos. It begins with hesitation, a quiet sense that something routine has stopped behaving predictably.

Failures spread in the background. Verification servers drift out of sync. Digital signatures fail to authenticate. Secure connections refuse to establish. Trust does not collapse with a dramatic outage. It dissolves as the systems that certify truth begin to doubt themselves.

The First Days of Chaos

The first major shock hits the financial world. Authentication systems reject valid logins. Banking apps display account balances that back-end databases cannot confirm. ATMs print receipts with timestamps that do not match reality. Some print dates from last week. Some show dates are from next month. Cash stops dispensing, but the machines remain powered. The network behind them no longer believes its own records, and neither do the people relying on it.

Cryptocurrency markets unravel next. Quantum-powered attacks forge private keys with precision, undermining the foundational belief that wallet ownership can be verified. Transactions flood the network with counterfeit legitimacy. Miners halt. Exchanges freeze. Value evaporates because confidence evaporates.

Critical infrastructure begins to wobble, not from loss of power, but from loss of confidence. Electrical grid operators receive control signals signed with certificates that no longer validate. Substations stay energized, but automated load-balancing systems hesitate, unsure which commands are authentic. Operators revert to manual procedures designed for emergencies, slowing response times across entire regions. The grid holds, but only because humans step in where machines no longer agree.

Water utilities encounter similar uncertainty. Treatment facilities continue operating, but telemetry feeds report conflicting chemical levels. Remote valves reject commands or execute them twice. Engineers must verify readings through physical sampling because digital instrumentation can no longer be relied upon to prove its integrity. Water still flows, but assurance disappears.

Traffic systems degrade quietly. Signal controllers accept timing updates that later fail verification. Lights fall out of synchronization across corridors. Autonomous traffic management systems disengage and revert to fixed cycles. Congestion spreads, not from accidents, but from hesitation. Navigation systems disagree on routing. Traffic slows to a crawl because the infrastructure can no longer confirm its own instructions.

Nuclear facilities do not melt down. They shut down. Safety systems do exactly what they were designed to do when trust erodes. Digital signatures protecting control logic fail validation. Automated systems trip into safe states. Plants disconnect from the grid out of caution, not catastrophe. The danger is not an explosion. It is a sudden loss of capacity layered onto an already stressed energy network.

Airline and airport operations do not fail catastrophically, but they slow under uncertainty. Aircraft remain flyable, yet trusted updates, flight plans, maintenance releases, and traffic coordination cannot be validated with confidence. Air traffic control increases digital separation as surveillance feeds and flight data lose a clear source of authority. Airports revert to manual identity and safety checks, which stretch turn times and ground aircraft that cannot be digitally cleared. Navigation and timing systems fall back to conservative modes, reducing capacity in already-constrained airspace. The risk is not collision, but congestion driven by doubt. Air travel remains safe, but brittle, as safety margins are consumed to compensate for lost trust.

Retail systems collapse soon after. Gas pumps reject every card. Grocery store terminals decline all payments. Customers walk away from full carts. The power stays on. The shelves stay stocked, but the link between commerce and trust fails.

Healthcare systems remain operational but trapped in uncertainty. Patient records load with mismatched signatures. Diagnostic tools produce conflicting results. Biometric checks fail without explanation. A physician attempts to retrieve a medical file and receives a warning that the source cannot be verified. Devices meant to provide clarity now produce unease. Treatment becomes a matter of judgment rather than certainty.

Elections descend into suspicion. Deepfake candidates appear in simultaneous livestreams, each claiming victory with perfect realism. Government websites load with invalid certificates. Ballot systems display valid signatures on altered entries. Officials shift to paper records not out of nostalgia but necessity. Democracy falters not through force but through doubt.

Not every failure in this moment originates from a cryptographic flaw. Some systems still function mathematically, but fail socially. Artificial intelligence exploits familiarity, authority, and urgency, causing people to distrust accurate signals while accepting fabricated ones. In these cases, perception collapses before verification, accelerating confusion even when encryption has not yet failed.

Communication platforms collapse last. Verified social media accounts post statements their owners never wrote. WhatsApp drops calls as encryption checks fail. Group chats fill with cloned messages sent by synthetic voices. Families argue over which version of a message is real. LinkedIn becomes a platform for counterfeit corporate announcements. Instagram influencers vanish and return as flawless synthetic copies. The forgeries look correct because they follow the same rules the originals once used.

Email failure completes the collapse. Outlook and Gmail list their own certificates as untrusted. Messages fail to sync. Attachments cannot be authenticated. Hospitals cannot share records. Executives cannot approve deals. Families cannot confirm a single sentence. When email falls, digital identity loses its anchor.

At home, the collapse becomes personal. Smart locks refuse access. Thermostats reboot endlessly. Cloud archives reject credentials. Navigation apps lose reliability as location data falls out of sync. Basic tasks falter. The world remains powered, illuminated, and online, but its sense of certainty disappears.

Governments declare emergencies as communication networks fail. Cash becomes essential. Barter returns in rural towns. Hospitals revert to manual processes. Conspiracies spread without a trusted channel to challenge them. Citizens lose confidence in digital institutions long before systems return online. The Zero Epoch begins not with destruction but with disorientation.

What this looks like in daily life is not darkness or immediate destruction, but progressive paralysis. Money exists, yet cannot be reliably accessed or exchanged. Power, water, transportation, and healthcare continue to operate, but under manual control and with constant doubt. Payments halt. Supply chains slow. Travel contracts and commutes are crippled. Systems meant to ensure safety become cautious rather than responsive. At home, locks, navigation, and basic services behave unpredictably. In public, information fragments. Elections are questioned. Official guidance competes with flawless fabrications. As trust erodes, coordination breaks down. Commerce pauses. Decision-making slows. When systems no longer agree, and people can no longer verify authority, society does not adapt smoothly. It begins to stall. The Zero Epoch does not announce itself through collapse. It arrives when daily life can no longer move forward with confidence.

The Inward Target

For decades, trust in digital systems has been treated as absolute. Machines were assumed to be neutral, deterministic, and incorruptible. When something went wrong, the conclusion was almost automatic. The system did not fail. A human must have intervened. Someone with privileged access must have abused it.

That assumption reshapes blame. Errors are not seen as emergent behavior or systemic weakness. They are treated as intent. Fraud implies an insider. Inconsistency implies sabotage. When logs cannot be trusted, and signatures no longer prove origin, suspicion defaults to the people closest to the controls, especially when there are no alerts.

In the Zero Epoch, that reflex becomes a tool of conquest orchestrated by invisible bad actors. Systems can lie convincingly without human direction. Cryptography can fail without malice. Automation can propagate false states at machine speed. Yet the instinct remains to look inward with prejudice. Teams scrutinize each other. Access becomes guilt. Privilege becomes motive.

The tragedy is that this suspicion feels rational. Humans have always been the weakest link, so when trust collapses, humans become the explanation. The result is isolation, accusation, and internal fracture at the exact moment coordination is most needed. When digital systems lose their claim to objectivity, trust does not disappear. It turns inward and starts consuming the people who built and maintained them.

Voices from the Blackout

A bank teller studies a transfer stamped with yesterday's date and wonders whether the system failed or whether someone with access altered it deliberately. A hospital CISO recognizes that an attack was designed to preserve data while destroying confidence in it, and begins to ask which administrators still had valid credentials when the signatures stopped validating. An election official prints hand-counted results and quietly inventories who had physical access to the machines before the outage.

A small business owner watches card readers fail and suspects the outage is selective, aimed at certain merchants, certain regions, certain people. A social media manager sees forged press releases appear on a verified account and wonders whether the breach came from outside the company or from someone who already knew the approval process. An engineer traces the failure to a single node that injected false state information and asks the question no one wants to ask first. Who had the authority to make it lie?

No one shouts, no one riots. The damage is subtler. Trust turns inward and sharpens. Colleagues begin to review each other's access histories. Longstanding permissions feel reckless. Every unexplained action raises the possibility of betrayal. In the absence of reliable proof, insider threat becomes the default explanation because it feels more controllable than randomness.

Fear does not arrive as panic. It arrives as vigilance, which hardens into suspicion. Suspicion isolates teams that once depended on each other. When systems cannot prove integrity, people look for intent, and intent is easiest to assign to someone already inside the walls.

The Lesson of Maple Street

Long before quantum computing or artificial intelligence, the Zero Epoch has already been rehearsed. In the Twilight Zone episode *The Monsters Are Due on*

Maple Street (season 1, episode 22), the threat is not invasion or destruction. It is the collection of odd things happening all at once. Street and house lights flicker. Power cuts in and out. Familiar systems stop behaving predictably. No explanation is offered, so fear fills the gap.

As trust erodes, neighbors turn on one another. Conspiracy replaces evidence. The collapse is psychological before it is physical. The aliens barely need to act. They only disrupt the signals people rely on to confirm reality. Society fractures and collapses on its own.

Paraphrased, the best way to conquer humans is not through force, but by turning them against each other. All that is required is to introduce fear, confusion, and distrust. Once that happens, humans will destroy themselves.

The actual dialogue, spoken during the closing narration and alien reveal, conveys that they do not need weapons. They place a neighborhood into darkness, disrupt a few systems, and let suspicion do the rest. The darkness is not just the absence of light. It is the absence of certainty.

That line matters because the aliens describe Maple Street as a model, not an exception. Any place can be pushed into the dark if people stop trusting shared signals and start assigning malicious intent to one another. That is precisely the failure mode described by the Zero Epoch.

The Zero Epoch anticipates the same pattern, but on a planetary scale. Financial records contradict each other. Certificates fail validation. Official statements become indistinguishable from perfect fabrications. When verification fails, conspiracy becomes easier than patience, and blame becomes easier than proof. It's only human nature that violence replaces reason.

Reflection

Midnight of the Zero Epoch arrives without noise. Screens glow. Networks hum. Systems respond somewhat as expected. What vanishes is certainty. The belief that a click implies consent, that a signature proves truth, that automation guarantees safety. Technology does not collapse in that moment. Belief does. The failure exposed is not in code or machines, but in the assumption that complexity could operate indefinitely without scrutiny.

As that belief fractures, watch as behavior changes. Individuals hesitate, but groups react. People pause before approving actions, but crowds move on fear rather than judgment. Signals are no longer accepted at face value, and fear spreads faster than explanation. This is where danger emerges. Not from broken systems, but from human response when trusted anchors vanish. Fear in isolation can lead to caution. Fear in groups turns outward. Suspicion replaces verification. Authority dissolves. Violence becomes a way to restore control when proof no longer exists.

"A person is smart. People are dumb, panicky, dangerous animals, and you know it." – Agent K, MIB

The pause that follows is not weakness. It is recognition of threat. Trust exposes its true nature in that moment. Unlike Y2K, it was never embedded in software. It was a promise people extended to systems they believed were stable, impartial, and honest. When that promise breaks, restraint weakens. Fear seeks certainty. Groups look for someone to blame. When fear dominates, a terrified society stops reasoning and starts attacking. It leaps, fragments, and risks turning that fear into violence against itself.

Rebuilding must begin in that fragile moment, but only if courage replaces reflex. Not the speed to act first, but the discipline to pause, to stop assigning blame to coworkers, and to see through conspiracies that offer comfort instead of truth. Recovery does not come from patches alone. It requires resolve to strengthen systems without turning on the people who maintain them.

The Zero Epoch becomes a hard dividing line. On one side is Maple Street, where disruption is misunderstood, fear fills the silence, trust collapses, and people turn on one another until destruction feels justified. On the other side is recognition, where disruption is expected, deception is assumed, and panic is refused. Trust is no longer granted by default. It is rebuilt deliberately through verification, transparency, and security postures hardened for a world where signals can lie. One path ends in a collapse. The other demands preparation, restraint, and the will to hold the line when fear insists otherwise.

The future is shaped not by failing machines but by whether humanity can master fear, resist panic, and commit to rebuilding trust and security together, step by deliberate step.

Chapter 10: Rebuilding Trust

The Turning Point

The collapse does not end the story. It begins the work that matters. When systems fail at the identity layer, the recovery is not a technical task. It is a reconstruction of confidence. Organizations must rebuild trust one verification at a time. Citizens must learn which signals remain reliable. Governments must establish credibility in a world where certainty cannot be assumed.

Recovery in the post-quantum age is not restoration. It is reconstruction. The systems that broke cannot return to their previous state. The rules have changed. The assumptions have changed. The threat surface has changed. Rebuilding trust requires new foundations.

Stabilizing the Aftermath

Chapter 10: Rebuilding Trust

In the first hours after a trust failure, priority shifts from precision to survival. Organizations must triage the bleeding before they understand the cause. The first step is containment. The second is fidelity.

Teams isolate compromised identity systems. They suspend automated decisions. They cut access to external partners. They freeze all cryptographic functions that depend on broken mathematics. They move to manual operation for anything related to authentication, authorization, or signature validation.

This phase is slow and uncomfortable. It requires people to do what machines once did. It introduces delay into systems built for speed. The goal is not efficiency. The goal is control.

Restoring Command and Communication

When trust collapses, communication becomes the first casualty. Leaders cannot rely on email, messaging platforms, digital signatures, or identity tokens. Every channel that claims authenticity becomes suspect. Messages that once felt routine become potential attack vectors. Instructions that once flowed freely now carry risk.

Recovery begins by shifting authority away from compromised systems and toward channels that cannot be forged. These channels do not depend on cryptographic confidence. They rely on physical control. They restore leadership's ability to speak with clarity and to know that the message received is the message sent.

Organizations activate secure out-of-band routes. They move to dedicated phone lines that bypass enterprise networks. They conduct in-person briefings for decisions that shape operational posture. They rely on paper-based confirmations for approvals that cannot tolerate doubt. They store sensitive materials in controlled spaces where access can be verified by sight, not by software.

These channels are intentionally slow and disruptive to everyday workflow. The friction is deliberate. It forces human judgment to replace automated trust at critical moments. By slowing automation, these channels prevent attackers from injecting false instructions into command processes. They also reduce the risk that confusion or miscommunication turns a technical failure into an organizational failure.

Once leadership reclaims a trusted channel, it can coordinate the response. Teams can receive verified orders. Departments can align on priorities. External partners can confirm status without relying on compromised digital paths. Communication becomes deliberate rather than assumed.

Only after these channels stabilize does the organization move to the next stage of recovery. Systems can be rebuilt. Keys can be reissued. Workflows can be restored. None of these efforts succeed if command, control, and communication remain in doubt.

In the post-quantum age, restoring communication is not a procedural step. It is the anchor for every action that follows.

Verifying the Ground Truth

Reconstruction begins with a single objective. Identify what is real. After a trust collapse, organizations cannot assume that any digital signal is accurate. Systems that once verified themselves through cryptography no longer provide certainty. The guarantees that protected

Chapter 10: Rebuilding Trust

identity, data, and workflow must be recreated from first principles.

Teams begin by isolating every system touched by compromised keys or broken algorithms. They separate networks by trust level. They tag data sources as confirmed, suspected, or unknown. They treat every unexplained change as potential manipulation. The goal is not speed. The goal is clarity.

Verification moves from digital assurance to physical confirmation.
- Servers are inspected.
- Logs are examined by hand.
- Audit trails are matched against external evidence.
- Trusted personnel confirm the provenance of critical decisions.
- Independent teams cross-check findings to avoid blind spots.

This work is methodical by necessity. It forces organizations to reconstruct an accurate view of their own environment after disruption. Teams must confirm which systems continued to operate correctly, which deviated from expected behavior, and which kept processing data even after trust mechanisms failed. This clarity is required before recovery can begin.

Identity must be rebuilt before anything else.
- Teams revoke compromised credentials.
- They issue new credentials using post-quantum algorithms.
- They verify every identity through physical or in-person checks.
- They document each step so future audits can validate the chain of custody.

Data recovery follows the same pattern.

- Teams identify authoritative records from offline backups.
- They compare them against live systems.
- They remove corrupted entries.
- They rebuild datasets from copies stored outside the attack surface.
- They track discrepancies to determine whether they resulted from failure, manipulation, or drift.

Process verification closes the loop.
- Teams map expected workflows.
- They watch how systems behave under controlled conditions.
- They confirm that outcomes match intentions.
- They correct any divergence before reconnecting the process to broader systems.

This work consumes time and manpower. It slows operations and strains resources. Yet it is the only way to establish a new baseline that organizations can trust. Without these deliberate steps, every subsequent action rests on compromised assumptions.

Ground truth forms the foundation of recovery. When trust breaks down, certainty needs to be reconstructed gradually, piece by piece.

Post-quantum Incident Response

Incident response in the new age requires a fundamental shift in mindset. Traditional models assume that attacks leave traces. They assume that logs provide sequences. They assume that identities fail in visible ways. Quantum-enabled attacks challenge every one of those assumptions.

A quantum-capable adversary can decrypt information without touching the system that holds it. They can extract secrets while leaving no artifacts

Chapter 10: Rebuilding Trust

behind. They can alter data without breaking signatures because the signatures themselves have lost meaning. They can bypass audit trails that rely on cryptographic integrity. These attacks leave no footprint that a classical response team is trained to recognize.

Post-quantum incident response cannot depend on the artifacts that guided earlier investigations. It must focus on patterns, not proofs. Analysts search for behavioral anomalies rather than cryptographic inconsistencies. They look for unexpected relationships in data flows. They study timing, frequency, and system load rather than trust certificates or tokens. They evaluate what a system produces rather than what a system claims.

This approach treats every digital signal as unverified until confirmed through independent observation. It replaces automated certainty with human judgment. It transforms incident response from a forensic timeline into a behavioral analysis.

Teams recreate expected workflows and compare them with observed activity. They examine whether outcomes align with historical baselines. They correlate findings across multiple systems because a single data point no longer provides confidence. They use independent logs, physical records, and external partners to reconstruct events.

This model is slower. It is deliberate. It requires more attention and more people. Yet it is necessary. Quantum disruption removes the shortcuts that incident response once depended on. Recovery depends on understanding what happened, not on trusting the signals that claim to describe it.

In the post-quantum age, incident response becomes an exercise in discernment. It becomes a search for truth in a landscape where authenticity is no longer inherent. The goal is no longer to identify the attacker. The goal is to reestablish control over systems that may have been compromised without leaving any trace of intrusion.

Business Continuity in a Post-quantum Age

Business continuity plans must evolve. Modern organizations built their continuity around digital redundancy. They assumed that if one system failed, another system would take its place. They believed backups would remain clean, that signatures would remain valid, and that identity would remain stable. A post-quantum collapse breaks all of these assumptions at once.

Continuity can no longer rely on redundant systems that depend on the same broken cryptography. A mirrored server is not protection when both copies trust the duplicate vulnerable keys. A replicated database is not secure when each clone depends on signatures that no longer verify. Redundancy stops being a shield when every redundant system inherits the same weakness.

Continuity must include parallel operations that function without trust in digital identity. Organizations must be prepared to operate in a degraded mode that does not depend on automated validation or cryptographic guarantees. This requires a return to fundamentals.

Leaders prepare analog contingencies for critical decisions.
- They reintroduce manual approval chains.
- They designate individuals who hold authority offline.

- They store essential records in controlled physical environments.
- They maintain printed procedures for emergency use.

Financial operations receive the same treatment.
- Paper-based workflows support transactions that cannot risk digital compromise.
- Checks, vouchers, and manual reconciliations become part of the fallback toolkit.
- Offline ledgers provide continuity when digital integrity fails.

Staff require training for these modes.
- They learn to verify identity without digital credentials.
- They learn to validate information through cross-checks instead of automated trust.
- They learn to perform tasks normally delegated to machines.

Organizations must design fallback modes that reduce capacity but maintain integrity. Systems must be able to run with limited features to avoid trust-dependent functions. Networks should operate with restricted access to minimize attack vectors. Digital processes continue only where post-quantum protections are in place.

Continuity becomes a spectrum rather than a switch.
- Full automation in normal conditions.
- Hybrid operation when disruption creates uncertainty.
- Human-centered control when digital identity cannot be trusted.

This layered approach maintains function even when the digital core becomes unreliable. It enables organizations to continue operating during uncertainty.

It buys time for recovery teams to rebuild trust at the pace that the new threat environment demands.

In the post-quantum age, continuity is no longer about preventing outage. It is about preserving integrity when the systems designed to guarantee integrity become the point of failure.

Rebuilding Identity at Scale

Identity is the backbone of recovery. When quantum computing breaks classical encryption, every credential tied to that math becomes suspect. Passwords, private keys, digital certificates, and token-based systems all lose their assurance. Organizations cannot rebuild trust until they rebuild identity. They must transition to post-quantum identity systems that support encryption, signatures, and authentication that are resilient to quantum attacks.

The transition begins by revoking compromised keys. This step removes the false sense of security that broken credentials provide. Every key that relies on classical algorithms must be treated as unsafe. Revocation cascades across users, devices, applications, and partners. It forces a reset of the entire identity landscape.

Next, organizations issue new keys based on quantum-resistant algorithms. These include post-quantum signatures and key establishment methods selected for their resilience. This step requires careful deployment. Key material must be stored in hardened modules. Issuance must be verified through physical or multi-person checks. Every identity must be reconstructed with a chain of custody that cannot be forged.

The third step requires alignment across the ecosystem. Partners, vendors, cloud providers, contractors, and service integrators must adopt the same post-quantum standards. Identity cannot be secure if the organization trusts credentials issued by entities still using broken cryptography. Negotiation, coordination, and enforcement become essential.

The fourth step implements monitoring to detect attempts to exploit legacy systems. Attackers will search for outdated applications, forgotten certificates, and old configurations. Continuous scanning and configuration validation become core components of identity management. Any reversion to classical algorithms becomes a security incident.

Identity recovery succeeds only when the entire ecosystem adopts the new model. Partial migration leaves gaps wide enough for attackers to operate without detection. Identity is no longer a local decision. It is a collective responsibility.

Rebuilding Institutional Credibility

Trust does not return when the systems return. It returns when institutions prove they understand why the failure occurred and show how they will prevent the next one. Systems restore capability. Leadership restores confidence.

Leaders must communicate clearly. They must lay out what failed, how it failed, and what the organization learned. They must show their decisions. They must articulate the tradeoffs they faced. They must explain their constraints without using them as excuses. They must provide evidence of progress rather than promises.

Competence under pressure becomes a greater signal than stability under normal conditions. People watch for discipline. They watch for consistency. They watch to see whether leaders continue the work after public attention fades. The rebuilding of credibility is not a press release. It is a pattern of behavior.

Citizens and customers respond to transparency and discipline. They do not respond to reassurance alone. Confidence grows when actions match commitments. Institutions must show that they learned from failure, not that they waited for conditions to improve.

Rebuilding credibility is not optional. It determines whether people trust the next warning, the next patch, or the next emergency directive. Without credibility, even correct information fails to persuade.

Rebuilding Social Confidence

Society will need to rebuild trust, but that trust can no longer rely on signals that machines can fabricate. People must adopt deliberate habits of verification. Familiar voices, known faces, and polished documents must be questioned rather than accepted by default. The critical skill will be learning to distinguish fast information from information that has been confirmed.

This transition requires education. Not technical training, but awareness training. People need to understand the limits of AI and the fragility of digital certainty. They must know which signals are stable and which signals are vulnerable. They must become comfortable with delayed confirmation and slower certainty.

Families, communities, workplaces, and schools must encourage practices that reduce impulsive trust. People must learn to pause before forwarding a message. They must learn to cross-check sources. They must learn to identify anomalies in tone, language, and timing. These habits cannot be imposed. They must be cultivated.

Confidence needs to be rebuilt from the ground up rather than assumed from the top down. While institutions can restore credibility, it is the citizens who must rebuild trust.

In the post-quantum age, the human ability to verify becomes as crucial as the machine's ability to compute.

Rebuilding the Architecture of Trust

The final stage is the reconstruction of the trust infrastructure itself. This is not a repair. It is a redesign. The old architecture relied on assumptions that no longer hold. It relied on cryptography that no longer protects. It relied on verification methods that machines can now imitate.

New trust begins with new math. Post-quantum algorithms must replace the classical algorithms that no longer resist attack. Encryption, signatures, and key exchange must be rebuilt on foundations selected for resilience against quantum capabilities.

This work must begin now.

Re-engineer hardware roots of trust today. Secure enclaves, firmware controls, and hardware-backed identity modules must move toward post-quantum protection paths. Physical components can no longer be treated as secondary to software. When cryptography weakens, hardware becomes a primary line of defense.

Rebuild verification frameworks with the assumption that deception is routine. Require confirmation from independent sources before proceeding with critical actions. Anchor logs in immutable storage. Insert human checkpoints where automated trust once ruled. Speed must yield to certainty when decisions carry real consequences.

Adopt crypto agility as a design requirement, not a future upgrade. Build systems that can change algorithms, rotate keys, and migrate trust mechanisms without collapse. Retire static standards that assume long-term stability. Flexibility must replace permanence.

Design systems to expect failure, not stability. Components should degrade safely, isolate compromise, and shift into safe modes rather than halt entirely. Continuity matters more than perfection when disruption becomes unavoidable.

Move audits into continuous operation. Stop treating oversight as a quarterly or annual event. Monitor every change, configuration shift, and identity action in real time. Visibility must exist while correction is still possible.

Make risk visible through behavior, not dashboards alone. Embed validation into daily operations. Shape processes around uncertainty instead of reacting to surprises. Prepared organizations show it in how they operate, not how they report.

This is the architecture required for a world where trust is no longer inherited by default. It must be earned, maintained, and verified continuously. The threat environment is changing now, shaped by quantum-capable adversaries and machines that can imitate signals once treated as proof. Preparation cannot wait for disruption to announce itself.

In the Zero Epoch, trust becomes a discipline. It is the foundation on which resilience is built.

Reflection

Rebuilding trust after the Zero Epoch is not a return to what once worked. It is an acceptance that the signals,

Chapter 10: Rebuilding Trust

certainties, and assumptions that shaped the last century no longer behave the same way. The world has changed. Trust must change with it.

Recovery requires discipline. Continuity requires resilience. Identity requires reinvention. Governance requires speed. Society requires awareness.

Each principle demands action. Each principle demands humility. Organizations must learn to rebuild trust from verified ground truth. Governments must learn to act before the threat matures. Citizens must learn to question the digital signals they once accepted without pause.

Trust will return, but slowly. It will return through verification, not tradition. It will return through transparency, not ritual. It will return through systems designed for a world in which machines operate at a pace humans cannot follow.

The collapse of trust reveals something more profound. The most serious risk was never the technology. It was the belief that technology did not need scrutiny. It was the willingness to accept certainty without confirmation. It was the habit of granting authority to systems that rarely explained themselves.

From crisis comes clarity. The Zero Epoch becomes a dawn rather than a decline. It forces humanity to rebuild trust with intention. It forces institutions to design systems that earn confidence rather than inherit it. It forces leaders to recognize that trust is not a convenience. It is an obligation.

Strength does not come from trying to avoid disruption. It comes from preparing for it with intent and discipline. It comes from designing systems that can absorb failure without losing their purpose. Above all, it comes from recognizing that trust is not something you possess. It is something you practice, continuously and deliberately.

This chapter explored how these lessons shape the world that follows. It examined the new foundations of credibility and the responsibilities that come with them. The world was prepared for Y2K because it had a deadline. The Zero Epoch has no deadline, but it will be marked by quantum computing breaking RSA encryption. Any critical infrastructure still using legacy encryption will become vulnerable to attack. The future can be stable if the choices made now reflect the reality of the age we are entering, not the comfort of the age we are leaving.

Epilogue: Final Reflection

Before the Clock Strikes Zero

The world will not disappear. If it fails, it will fail in confusion, and it will stop making sense. The danger of the Zero Epoch rests not in the strength of quantum computation, but in the slowness of our response to it. History rarely announces its turning points. They arrive disguised as postponed patches, missed updates, and neglected audits. Empires once fell from pride. Today, they could fall from expired certificates.

Quantum supremacy is approaching. When it arrives, the ability to pierce secrets hidden behind encryption will no longer be theory. The question is not if it will happen. The question is when.

Every leader who funds security, every engineer who strengthens a protocol, and every citizen who pauses to verify a message contributes to the preservation of truth. Machines will not write the future. People will. The future belongs to those who remember why these systems were built in the first place.

Readiness is ethical as much as technical. It asks us to defend truth even when convenience makes falsehood easier.

Though this is excellent material for an "end of the world" thriller, especially for a director with creative license. However, if the Zero Epoch arrives without resistance, the collapse will not look cinematic with volcanoes erupting, planes falling out of the sky, or buildings collapsing. It will be quiet, volatile, and chaotic. The cascading effect will be undeniable and unlike anything we've ever seen. Will it send us back to the dark ages? Doubtful, but in the hands of bad actors, this could be a different world.

This warning is prophecy. The Zero Epoch will occur, but its effects can still be mitigated.

Strengthening the Foundations: The Role of a Mature Cybersecurity Program

The Zero Epoch reveals how trust fails when identity weakens, when cryptography ages, and when adversaries exploit speed and ambiguity. Yet even in this environment, one principle remains constant. Strong cybersecurity programs create stability. They reduce the attack surface. They slow exploitation. They give organizations the structure they need to adopt post-quantum protections without improvisation.

A mature program uses disciplined controls and verified processes. These controls appear across many global frameworks. The NIST Risk Management Framework, the NIST Cybersecurity Framework, Cybersecurity Maturity Model Certification, ISO 27001, Zero Trust, and other standards exist for a reason. They require organizations to document, measure, and review the safeguards that keep systems stable. Frameworks create consistency to ensure cyber-readiness at all times. They force teams to build security that can survive stress rather than security built around convenience.

These frameworks differ in scope, but they reinforce the same idea. Security depends on structure. Without structure, no amount of technical talent or budget prevents failure. In the Zero Epoch, structure becomes essential because quantum acceleration and AI deception amplify existing weaknesses.

12 Pillars of Cybersecurity

A helpful way to view the basic foundations of cybersecurity is through the following 12 pillars. Each pillar answers a simple question: What must remain strong when everything else strains under pressure?

1. **Disaster Recovery**
   Disaster recovery exists to prove that systems can return from failure, not to assume they will. Backups must be tested through real-world restoration, not trusted by policy alone. Offline and immutable copies protect against ransomware and systemic compromise. Redundancy must be designed so that a single failure does not cascade into all copies failing simultaneously. Recovery only matters if it works under pressure.

2. **Authentication**
   Authentication establishes who is requesting access at a specific moment in time. Stolen credentials lose

their usefulness when access depends on more than knowing a password, and when that access is time-bound and tightly controlled. Identity must be continuously verified, not inferred from past success. When authentication is weak, every downstream control becomes meaningless. Trust begins here and nowhere else.

3. **Authorization**
Authorization determines what an authenticated identity is allowed to do. Access must align tightly with role and current need. Excess privileges create silent attack paths that remain invisible until exploited. Regular review and revocation matter more than initial approval. The safest access is the access that was never granted.

4. **Encryption**
Encryption protects confidentiality and integrity while data moves and while it rests. Classical algorithms no longer offer long-term assurance against future computation. Systems must begin preparing for post-quantum methods before failure forces emergency migration. Encryption is not a feature, it is a dependency that must evolve as the threat environment changes.

5. **Vulnerability Management**
Vulnerability management turns known weaknesses into measurable risk reduction. Continuous scanning reveals exposure before attackers do. Patching and configuration control close doors that adversaries actively test. Most breaches begin with flaws that were already documented and ignored. Discipline here prevents preventable failure.

6. **Audit and Compliance**
Audit and compliance provide evidence that controls exist and operate as intended. Logs,

assessments, and independent reviews surface weaknesses while correction is still possible. Compliance alone does not guarantee security, but the absence of evidence guarantees blind spots. Early visibility reduces late-stage damage.

7. **Network Security**
Network security limits how far an attacker can move once inside. Segmentation enforces boundaries that prevent single compromises from becoming systemic failures. Traffic inspection reveals misuse and abnormal behavior. Containment depends on structure. Flat networks reward intrusion.

8. **Terminal Security**
Endpoints are the most common entry point because they sit closest to users. Laptops and mobile devices must enforce encryption, baseline configuration, and update discipline. A single compromised endpoint can bypass perimeter defenses entirely. Protecting terminals protects the rest of the environment by extension.

9. **Emergency Response**
Emergency responses assume that information will be incomplete and misleading. Plans must function without perfect identity, perfect telemetry, or full visibility. Teams must practice decision-making under uncertainty. Speed matters, but control matters more. A response plan that depends on ideal conditions will fail when conditions degrade.

10. **Container Security**
Containers accelerate deployment, but also compress risk. Images must be scanned before use, privileges must be tightly restricted, and runtime behavior must be continuously observed. A vulnerable container can quickly spread

compromise across modern architectures. This layer now carries the weight once held by servers and must be treated with equal seriousness.

11. API Security
APIs expose core functionality directly to other systems. Every call must be authenticated, authorized, and monitored. Patterns reveal misuse faster than static rules. Services should operate with the minimum permissions required. APIs are doors, not pipes, and must be guarded accordingly.

12. Third-Party Management
Third parties extend capability, but also extend exposure. Vendors must be assessed, contracts must define security obligations, and inherited risk must be tracked continuously. A partner's weakness becomes your incident. Supply chain security is not optional when trust is shared across organizations.

These pillars align with every major framework. RMF maps them to controls. CSF organizes them into core functions. The Cybersecurity Maturity Model Certification embeds them into maturity levels. ISO 27001 connects them to policy, governance, and continuous improvement. Zero Trust reinforces them by treating identity, access, and verification as conditions that must be confirmed at every step rather than assumed. None of these models exist in isolation. They reinforce the same truth. Strong programs reduce fragility. Weak programs magnify it.

A strong program does not stop the Zero Epoch. It prepares the organization to face it. When the math fails, the tools that remain are discipline, structure, and the ability to maintain function while systems shift to post-quantum protections. No framework prevents disruption, but each one gives leaders a straightforward

Epilogue: Final Reflection

method to measure readiness and to correct gaps before adversaries exploit them.

In a world shaped by quantum acceleration and AI deception, the foundations matter more than ever. They create the stability that gives organizations time to migrate, adapt, and recover. Without these foundations, even the best technical defenses fail under pressure.

A Call to Action

The Zero Epoch demands a collective awakening. Preparation isn't confined to a single department or specialty but is a shared responsibility among engineers, executives, policymakers, and citizens. Each group plays a role in shaping the future of digital trust.

For Engineers: Design for Endurance, Not Speed

The future depends more on discipline than on breakthroughs. Every algorithm and every line of code becomes part of a shared nervous system that others must rely on. What you build will be stressed, misused, and pushed beyond its original intent. Design with the expectation of error. Build so systems continue to function when assumptions fail.

Systems must fail safely, not dramatically. Decisions should be documented so that others can understand the intent long after the context is lost. Encryption must be designed to change without collapsing trust or breaking dependencies. Transparency is a requirement because hidden complexity becomes hidden risk. Question every dependency, library, and service you rely on today.

Assume your tools may one day be used in ways you did not intend. Attackers adapt faster than documentation ages. Preparation is an engineering responsibility, not a policy exercise. The measure of excellence is not how fast something ships. It is how calmly it behaves when the environment around it starts to break.

For Executives: Fund the Work No One Applauds

Budgets expose priorities with perfect accuracy. The work that draws little attention, maintenance, upgrades, testing, and recovery planning, protects more value than visible initiatives ever will. Stable systems come from sustained investment in foundations. Cryptographic modernization, resilience training, and recovery capability require funding before they feel urgent. When these areas are ignored, failure becomes expensive and public.

Reward prevention as deliberately as response. Teams that stop incidents before they occur rarely receive recognition because success looks like nothing happened. Cybersecurity does not live inside a single office or title. It emerges from daily decisions, incentives, and risk tolerance. Leadership sets that tone through what it funds and what it excuses.

Organizations that endure the Zero Epoch will not be the ones that reacted fastest. They will be the ones who invested early and consistently. Resilience reflects leadership character. It shows whether continuity matters more to you than short-term comfort.

For Policymakers: Legislate for Transparency, Not Control

Power once came from secrecy. In the Zero Epoch, it will come from credibility. Write laws that protect truth. Require vendors to disclose their cryptographic practices. Demand independent verification for AI systems used in public life.

Build international interconnected frameworks for global post-quantum security. Fund education that teaches digital literacy as a civic skill. Governance must evolve quickly. The goal is not to restrain technology, but to ensure accountability within it.

A transparent government becomes a stable government. Control without clarity invites collapse.

For Citizens: The Everyday Defense of Truth

Citizens stand on the front line of the digital age. The Zero Epoch will begin when people trust a message that feels urgent, a screenshot that looks convincing, or an online service that still depends on cryptography vulnerable to quantum attack. Awareness becomes protection.

Every person can reduce risk by staying informed about quantum computing trends, choosing online services that confirm their post-quantum readiness, enabling stronger authentication, and verifying information before sharing it. These habits slow the spread of false signals and help maintain truth in a world shaped by rapid change. Your decisions create the first layer of defense long before any institution responds.

Personal vigilance scales faster than any firewall. Each informed user becomes a stabilizing force.

Practical steps include:
- Verify before sharing. Treat every claim as unconfirmed until checked across credible sources.
- Diversify information sources. Avoid relying on automated feeds alone.
- Protect digital identity. Use multifactor authentication, hardware keys, and password managers.

- Maintain offline copies of essential records.
- Support transparency laws and demand timely updates from service providers.
- Teach skepticism as a civic virtue. Normalize saying, "I am not sure. I need to check."

Security begins with perception. Awareness, curiosity, and restraint can outperform any algorithm. A population that verifies before acting becomes a collective firewall that cannot be bypassed.

The Final Reflection

Humanity's next frontier is not another planet or a virtual horizon. It is credibility itself. Modern life depends on the ability to trust what we see, hear, and exchange. Finance, healthcare, governance, and security all rest on that shared confidence. If we cannot preserve the truth that connects us, then no distance traveled and no technology invented will matter. Progress loses meaning when the signals that guide us cannot be believed.

Quantum disruption will not end humanity. It will stress the powerful undertones of human nature. It will expose whether people can guide the tools they create or whether those tools will outrun human judgment. Artificial intelligence will not decide our future. It will mirror our discipline, our priorities, and our willingness to question systems that appear reliable simply because they work.

The Zero Epoch is a prophecy, but not one of collapse. It is a moment of reckoning. It asks whether societies can rebuild trust for the world that exists, rather than cling to outdated models. It demands that identity, verification, and governance be treated as core infrastructure and anchor systems instead of being layered on later.

The future will not be shaped by algorithms alone, but by our choices today. By leaders who act before disruption becomes visible. By organizations that strengthen security as a daily practice, not as a reaction to incidents. By citizens who learn to question, verify, and critically think before accepting what appears real.

The Zero Epoch marks a seismic shift in the story of the digital world. It is a risk, but it is also an opportunity. Machines will not decide the future of digital trust. It will be defined by whether humanity chooses to act with clarity, awareness, and urgency while it still can.

Appendix: Zero Epoch Checklist

This checklist converts the concepts of The Zero Epoch Guidebook into a practical assessment. Each question probes readiness, resilience, and awareness. The objective is understanding rather than compliance, identifying where preparedness is solid and where it must be addressed.

Domain: Technological Readiness

Ask whether your systems can survive a loss of cryptographic certainty.

1. Do you know where encryption is used across your environment, not where you assume it is used?
2. Can you quickly identify which systems would fail first if certificates became untrusted?
3. Do you know which data must remain confidential for 10, 20, or 30 years?
4. Have you discussed post-quantum cryptography as a timing issue rather than a research topic?
5. Can your systems replace cryptographic algorithms without a complete redesign?
6. Do you rely on third parties to manage cryptographic decisions you do not fully understand?
7. Have you tested what happens when identity validation suddenly slows or fails?
8. Do your backups assume cryptographic trust during restoration?
9. Can your incident response operate if signatures and logs cannot be trusted?
10. Would you recognize a cryptographic failure before customers or citizens do?

Domain: Organizational Readiness
Ask whether leadership is prepared to act before certainty disappears.

1. Has executive leadership discussed quantum risk without waiting for a deadline?
2. Is there a clear owner for post-quantum transition decisions?
3. Do strategy discussions treat cryptography as infrastructure, not tooling?
4. Would your organization invest now to prevent a failure no one can yet see?
5. Do procurement decisions consider long-term cryptographic survivability?
6. Are supply chain partners expected to explain their cryptographic posture?
7. Does governance move fast enough to respond to emerging cryptographic risk?
8. Are risk discussions framed around trust failure, not breach counts?
9. Could your organization function temporarily with slower, manual verification?
10. Would leadership recognize hesitation and doubt as early warning signals?

Domain: Human Readiness
Ask whether people can function when digital trust weakens.

1. Do employees understand what encryption actually protects in daily work?
2. Would staff pause when a message looks right but feels wrong?

3. Are people trained to question authority signals delivered digitally?
4. Do leaders expect hesitation to increase during trust disruption?
5. Can teams operate effectively without instant verification?
6. Are fatigue and overload treated as security risks?
7. Do users understand that valid-looking does not mean authentic?
8. Are people encouraged to slow down when systems behave unpredictably?
9. Would employees know how to escalate uncertainty without fear?
10. Does your culture reward caution as much as speed?

www.ingramcontent.com/pod-product-compliance
Lightning Source LLC
LaVergne TN
LVHW010331070526
838199LV00065B/5719